# Social Computing
# and
# Behavioral Modeling

T0138048

# Social Computing
# and
# Behavioral Modeling

Edited by

Huan Liu

John J. Salerno

Michael J. Young

 Springer

*Editors:*

Huan Liu
Ira A. Fulton School of Engineering
Arizona State University
Brickyard Suite 501, 699 South Mill Avenue
Box 878809
Tempe, AZ 85287-8809, U.S.A.
huanliu@asu.edu

John J. Salerno
United States Air Force Research
    Laboratory
525 Brooks Road
Rome, NY 13441, U.S.A.
John.Salerno@rl.af.mil

Michael J. Young
Air Force Research Laboratory
2255 H Street
Wright-Patterson AFB, OH 45433-7022, U.S.A.
michael.young@wpafb.af.mil

ISBN-13: 978-1-4419-5491-6          e-ISBN-13: 978-1-4419-0056-2
DOI: 10.1007/978-1-4419-0056-2

Printed on acid-free paper.

© Springer Science+Business Media, LLC 2010

9 8 7 6 5 4 3 2 1

springer.com

# Preface

Social computing is concerned with the study of social behavior and social context based on computational systems. Behavioral modeling reproduces the social behavior, and allows for experimenting, scenario planning, and deep understanding of behavior, patterns, and potential outcomes. The pervasive use of computer and Internet technologies provides an unprecedented environment of various social activities. Social computing facilitates behavioral modeling in model building, analysis, pattern mining, and prediction. Numerous interdisciplinary and interdependent systems are created and used to represent the various social and physical systems for investigating the interactions between groups, communities, or nation-states. This requires joint efforts to take advantage of the state-of-the-art research from multiple disciplines, social computing, and behavioral modeling in order to document lessons learned and develop novel theories, experiments, and methodologies in terms of social, physical, psychological, and governmental mechanisms. The goal is to enable us to experiment, create, and recreate an operational environment with a better understanding of the contributions from each individual discipline, forging joint interdisciplinary efforts.

This is the second international workshop on *Social Computing, Behavioral Modeling and Prediction*. The submissions were from Asia, Australia, Europe, and America. Since SBP09 is a single-track workshop, we could not accept all the good submissions. The accepted papers cover a wide range of interesting topics. After we collected keywords from the papers, we made a naive attempt to group them based on their similarity into some themes: (i) *social network analysis* such as social computation, complex networks, virtual organization, and information diffusion; (ii) *modeling* including cultural modeling, statistical modeling, predictive modeling, cognitive modeling, and validation process; (iii) *machine learning and data mining*, link prediction, Bayesian inference, information extraction, information aggregation, soft information, content analysis, tag recommendation, and Web monitoring; (iv) *social behaviors* like large-scale agent-based simulations, group interaction and collaboration, interventions, human terrain, altruism, violent intent, and emergent behavior; (v) *public health* such as alcohol abuse, disease networks, pandemic influenza, and extreme events; (vi) *cultural aspects* like cultural consensus, coherence, psycho-cultural situation awareness, cultural patterns and representation, population beliefs, evolutionary economics, biological rationality, perceived trustworthiness, and relative preferences; and (vii) *effects and search*, for example, temporal effects, geospatial effects, coordinated search, and stochastic search. It is interesting that if we traced these keywords back to the papers, we could find natural groups of authors of different papers attacking similar problems.

We are extremely delighted that the technical program encompasses keynote speeches, invited talks, and high quality contributions from multiple disciplines. We warmly welcome all to actively participate in the interdisciplinary endeavors, and truly hope that our collaborative, exploratory research can advance the emerging field of social computing.

Phoenix, Arizona,                  *Huan Liu, John Salerno, and Michael Young*
                                           March 31 - April 1, 2009

# Acknowledgements

We would like to first express our gratitude to all the authors for contributing an extensive range of research topics showcasing many interesting research activities and pressing issues. The regret is ours that due to the space limit, we could not include as many papers as we wished. We thank the program committee members for helping review and provide constructive comments and suggestions. Their objective reviews significantly improved the overall quality and content of the papers. We would like to thank our keynote and invited speakers by presenting their unique research and views. We deeply thank the members of the organizing committee for helping to run the workshop smoothly; from the call for papers, the website development and update, to proceedings production and registration. We also received kind help from Kathleen Fretwell, Helen Burns, Connie Mudd, and Audrey Avant of the School of Computing and Informatics, ASU in many tedious but important matters that are transparent to many of us. We truly appreciate the timely help and assistance provided by Melissa Fearon and Valerie Schofield of Springer US to make the publication of the proceedings possible in a relatively short time frame. Working with ACM and SIGKDD (thanks to Gregory Piatetsky-Shapiro), SPB09 is in-cooperation with ACM SIGKDD. Last but not least, we sincerely appreciate the support from AFOSR, AFRL, ONR, NIH, and NSF.

We thank all for kind help and support to make SBP09 possible.

# Contents

# List of Contributors

Denise Anthony, Dartmouth College
Elsa Augustenborg, Pacific Northwest National Laboratory
William Batchelder, University of California Irvine
Nadya Belov, Lockheed Martin
Luis Bettencourt, Los Alamos National Laboratory
Colin Camerer, California Institute of Technology
Kathleen M. Carley, Carnegie Melon University
Sun-ki Chai, University of Hawaii
John Crosscope, Applied Systems Intelligence, Inc.
George Cybenko, Dartmouth College- Thayer School of Engineering
Gary Danielson, Pacific Northwest National Laboratory
Anthony Ford, US Citizen
Norman Geddes, Applied Systems Intelligence, Inc.
Vadas Gintautas, Los Alamos National Laboratory
Rebecca Goolsby, ONR
Aric Hagberg, Los Alamos National Laboratory
Michael Hechter, Arizona State University
Tristan Henderson, University of St. Andrews
Shoji Hirano, Shimane University
Shuyuan Mary Ho, Syracuse University
Ruben Juarez, University of Hawaii
Nika Kabiri, University of Washington
Masahiro Kimura, Ryukoku University
James Kitts, Columbia University
Baoxin Li, Arizona State University
Michael K. Martin, Carnegie Melon University
Hiroshi Motota, Osaka University and AFOSR/AOARD
Alce Mulvehill, US Citizen
David Murillo, Arizona State University
Ryohei Nakano, Chubu University

Jeff Patti, Lockheed Martin
Angela Pawlowski, Lockheed Martin
Alex (Sandy) Pentland, MIT
Colleen Phillips, Applied Systems Intelligence, Inc.
Jeff Reminga, Carnegie Melon University
Bruce Rogers, Arizona State University
Kazumi Saito , Univesity of Shizuoka
Nils F Sandell, Dartmouth College Thayer School of Engineering
Antonio Sanfilippo, Pacific Northwest National Laboratory
Robert Savell, Dartmouth College- Thayer School of Engineering
Jack Schryver, Oak Ridge National Laboratory
Shade Shutters, Arizona State University
Stacey Sokoloff, Applied Systems Intelligence, Inc.
Tomomi Tanaka, Arizona State University
Sandy Thompson, Pacific Northwest National Laboratory
Shusaku Tsumoto, Shimane University
David Twardowski, Dartmouth College- Thayer School of Engineering
Zheshen Wang, Arizona State University
Paul Whitney, Pacific Northwest National Laboratory
Ronald Yager, Machine Intelligence Institute - Iona College

# Organizing Committee

Huan Liu, John Salerno, Michael Young, Co-Chairs
Nitin Agarwal, Proceedings Chair
Guozhu Dong, Poster Session Chair
Sun-Ki Chai, Social Science Area Chair
Lei Tang and Magdiel Galan, Publicity Chairs
Sai Moturu, Local Arrangement Chair

# Program Committee

Lada Adamic, University of Michigan
Edo Airoldi, Princeton University
Joseph Antonik, AFRL
Chitta Baral, ASU
Herb Bell, AFRL
Lashon Booker, MITRE
Sun-Ki Chai, University of Hawaii
Kathleen Carley, CMU
Hsinchun Chen, Univiversity of Arizona
David Cheung, University of Hong Kong
Gerald Fensterer, AFRL
Laurie Fenstermacher, AFRL
Tim Finin, UMBC
Harold Hawkins, ONR
Michael Hinman, AFRL
Hillol Kargupta, UMBC
Rebecca Goolsby, ONR
Anne Kao, Boeing
Masahiro Kimura, Ryukoku University, Japan
Alexander Levis, GMU
Huan Liu, ASU
Mitja Lustrek, IJS, SI

Dana Nau, University of Maryland
Alex (Sandy) Pentland, MIT
Kazumi Saito, University of Shizuoka, Japan
John Salerno, AFRL
Antonio Sanfilippo, PNNL
Hessam Sarjoughian, ASU
Olivia Sheng, Utah University
Jaideep Srivastava, UMN
H. Eugene Stanley, Boston University
Gary Strong, Johns Hopkins University
V.S. Subrahmanian, University of Maryland
John Tangney, ONR
Tom Taylor, ASU
Bhavani M. Thuraisingham, UT Dallas
Ray Trechter, Sandia National Laboratory
Belle Tseng, Yahoo! Research
Xintao Wu, UNCC
Ronald Yager, Iona College
Michael Young, AFRL
Lei Yu, Binghamton University
Philip S. Yu, UI Chicago
Jun Zhang, AFOSR

Terry Lyons, AFOSR
Patricia L. Mabry, NIH
James Mayfield, JUH
Hiroshi Motoda, AFOSR/AOARD

Jianping Zhang, MITRE
Ding Zhou, Facebook Inc.
Daniel Zeng, University of Arizona

# Reality Mining of Mobile Communications: Toward A New Deal On Data

**Alex Pentland**
pentland@mit.edu, **MIT Media Laboratory**

``Within just a few years "people data" will be 90% of the world's collective data.''
*Jeff Nick, CTO of EMC, personal communication*

**Abstract.** Humanity has the beginnings of a new nervous system – a digital one derived from mobile telephone networks and already nearly four billion people strong. Computational models based on these digital "people data," using a process called *reality mining*, allow us to create a startlingly comprehensive picture of our lives, predicting human social behavior with a power that was barely conceivable just a few years ago. This new "god's eye" view of humanity will present unprecedented opportunities for profitable commerce and effective government ... and it will endanger our personal freedom [1]. To harness the good and avoid the evil, we need to forge a `New Deal' about how personal data can be gathered and used [2].

# References

1. Pentland, A., (2008). *Honest Signals: how they shape our world*, MIT Press, Cambridge MA
2. Sense Networks, http://www.sensenetworks.com

H. Liu et al. (eds.), *Social Computing and Behavioral Modeling*,
DOI: 10.1007/978-1-4419-0056-2_1, © Springer Science + Business Media, LLC 2009

# Lifting Elephants: Twitter and Blogging in Global Perspective

Rebecca Goolsby[†], Ph.D.

[†] Office of Naval Research

**Abstract** When we think about the actions that people are taking as a group, we must think how they interact to form patterns of collective behavior, not just how each individual acts. The patterns that arise result from the structure of interactions between individuals and benefits can arise from both connections and disconnections between individuals. Connections can lead to similar or coordinated behavior, which is important when the task involved requires such coordination. Separation and a lack of communication are required when there are independent subtasks to perform [1].

## 1 Introduction

Absolutely no one is trying harder or more creatively to develop the cutting edge in social use for the latest technologies than teenagers. They cheat on tests using SMS; they stare at their cellphones during conversations with their parents, multi-tasking between their "real" lives and their family lives. They "twitter". They facebook. They myspace. They are constantly online, connected more to one another than any group of young people outside of an Israeli kibbutz.

At one point, my seventeen year old daughter began to experience connection fatigue. She had over two hundred contacts on her cell phone. She was continually dealing with people with whom she had only a casual interest, which took away from time with the list of friends with whom she did want to interact. The disconnection/separation phase took weeks to achieve. When she had it optimized, it because a simple matter to coordinate parties, spur of the moment gatherings, even a barbeque for fifty people, organized through one click of a button—in a period of two hours.

Social computing, the use of innovations in technology and software to connect to other people, learn more about other people, and change social behavior owing to the benefits of real-time and asynchronous information feeds, has been with us awhile. One could say it began with the creation of email, or perhaps a bit later, with the birth of the world-wide web, that innovation of Sir Tim Berners-Lee. Some of us hoary heads recall the thrill of the first command line chat functionality back in the mid-80s

H. Liu et al. (eds.), *Social Computing and Behavioral Modeling*,
DOI: 10.1007/978-1-4419-0056-2_2, © Springer Science + Business Media, LLC 2009

and the British MUDS where we slew monsters and on the sidelines, discussed personal lives. And if you know what TS [2] stands for, yes, you are probably old.

Today social computing is not about email, MUDDING Or MUSHING, or any of those venerable technological wonders that are commonplace enough to have dictionary entries in Webster's. Social computing-- blogging, texting, facebook/myspace and their ilk, twittering (also known as microblogging) and on and on—these technologies are changing the world.

Different cultures and societies are using social computing ways that are changing attitudes and worldviews in almost every part of the world, including Africa, the least wired continent.

Twitter is about immediacy; tweet streams consist of short messages to ephemeral groups of "followers" who may come and go at will. There is an "everybody channel" where you can see every tweet, a slow moving barrage of random messages in many languages and scripts. To get the true flavor of Twitter, one must "tweet" and "tweet" in communities. Tweet for your "followers"—those people who have chosen to monitor your stream (and if you in turn monitor their tweet stream, it's a conversation.) Twitter is primarily a youth phenomenon, connecting up social groups that may be distributed over a city or a continent—or across the globe. It is very low bandwidth; people can and do twitter from cell phones. People use twitter to keep others abreast of what's going on and to coordinate behaviors on the fly. Twitter is not "broadcast"—it is more "narrowcast", requiring would be audiences to locate the individual "tweeter" and subscribe to their stream. There are some communal "tweeters" or "group tweets" as well. Alternately there are various search methods to locate streams of interest.

Twitter is also extremely international. The extent of that, and its potential importance, was made shockingly clear in 2008 with the use of Twitter when it became a critical element in the Mumbai tragedy. Twitter users provided instant eyewitness accounts all over Mumbai [3].

*Many Twitter users sent tweets [messages] pleading for blood donors to make their way to the JJ Hospital in Mumbai, as stocks were in danger of running low in the wake of the atrocity. Others spread the word about helplines and contact numbers for those worried about loved ones caught up in the attacks.*

But Twitter is not for the uninitiated. It can produce entirely too much noise, as David Sarno writes [4]:

*"During the attacks, users from around the world posted tens, if not hundreds, of thousands of short notes, updates, musings and links to the latest information on Mumbai — many, if not most, of the facts coming from mainstream news outlets. . . Though it's certainly possible with the right amount of patience and know-how, finding useful "tweets" during a major event like this is a little like panning for gold .....in a raging river."*

The problem with a citizen news network is that it is just that, unlettered, prone to mistakes, low on caution. Tweets were therefore often wrong, misleading, or simply un-useful---and almost always unverified and unsourced. The problem of noise is a constant one, with tweets echoing the same pieces of information. Or providing wrong information, such as the Tweets that asserted there was an attack on the Marriot Hotel, when there was no such attack [5]. Still, when the situation involves many people directly observing the same situation directly, the "streams" (groups of messages) can be invaluable [6]. Twitter is also useful for finding out what people are thinking about an event (op.cit.), such tweets may be "noise" if you want hard facts, but very useful if you are trying to understand public opinion about those facts.

Claudine Beaumont, of the UK Telegraph reports that within minutes of the attack, Wikipedia had set up an entry with "a team of citizen editors adding a staggering amount of detail, often in real time, to provide background information about the attacks." Beaumont credits one of the most riveting set of photographs to Vinu Ranganathan , whose Flickr photos brought a stream of immediate, on the ground images of the attack to the world [7]. Because it is continually being edited, bad information was not repeated in the system for any length of time, a kind of implicit editing function. Opinion information was likewise jettisoned.

One of the main differences between tweets and blogs is length. Tweets are no more than three lines—there is a 140 character limit on tweets--blogs can go on for pages. Tweets can embed an URL pointer to something else on the web (such as a blog posting or webpage)—and they frequently do. Further, social networking sites like Facebook advertise Twitter. For example, the new South Africa Twitter project, SA Twitter, advertises on its Facebook page, another space where (predominately young) people hook up with those with common interests. The South African Twitter project has set up a kind of group object, SAPROJECT, a "mash up" of tweets all syndicated into a single feed (http://twitter.com/saproject) that allows for those interested in South Africa to create a kind of micro-blog forum.

Twitter is not just for kids. BBCWORLD and other news organizations, as well as political organizations like openDemocracy, maintain a Twitter presence. There is even a blog "Twitip" (http://www.twitip.com) dedicated to teaching people how to twitter.

The brainchild of Darren Rowse, Twitip provides a commentary of all things on Twitter. His blog provides an analysis of social behavior in Twitter, as well as advice. In a more recent guest post, Snow Vandermore notes that there are many people who don't "get" Twitter: people on Twitter who only post blog (after blog after blog) URLs, a tendency she calls "blog vomit".

*"If you don't invest some effort in following and conversing with likeminded Twitter members, you might as well be broadcasting to Mars, because no one will notice you."*

People who hype their online businesses, indulge in fantasy lives revolving around sex and excess, people posting sexy avatars in the hopes of hooking up compete with would-be comedians, lurkers, "town criers" (who post news stories in a flash), they fail to realize the social potential of establishing a prolonged conversation with other "Tweeple" (Rowse's word for Twitterers).

For Vandermore, Twitter is about agendas. Each of these Twitter types has an agenda that is overly focused and one dimensional. The happiest Twitterers have more than one agenda. She writes:

*[When you have multiple agendas you can be] [v]ery efficient — managing to pimp your blog or website, spread viral headlines and videos, and flirt — in one perfect 140 character tweet (also referred to as a twoosh.) Barack Obama did it and so can you — achieve total Twitter nirvana [8].*

Vandermore writes that the longer people frequently Twitter, the more likely they will find themselves developing more than one agenda. Such multi-taskers realize that one can kill two or three birds with one tweet.

In the developing world, there are also many of these agendas. In Africa and Asia, (as well as Western nations) Twitter fills the need for a small bandwidth coordination and collaboration platform that is implicitly mobile. Twitter's capability to broadcast URLs on the fly, from cellphones, allowing Twitterers to cover fast breaking events like the Mumbai attacks. Their ability to direct people to blogs and news feeds covering current events  provides for a coordinated audience of concerned people, who are not isolated, but can trade comments and discuss.  Further, if these people regularly Twitter, they become known to each other, building bonds of trust.  The potential for coordinated action becomes increasingly real.

Blogs in the underdeveloped world have come of age in the last three to five years, particularly in Africa. They have covered crucial events, such as violence during Kenyan elections and a large number of grassroots movements, from an anti-smoking campaign to poverty issues to transparency and corruption in the Kenyan parliament [9]. They are of increasing concern to corrupt governments that many of these countries are putting pressure on bloggers. Egypt, Morocco and Nigeria have reportedly harassed bloggers, arresting (and then releasing them) on spurious charges [10].  Mumbai has shown the potential for using microblogging systems like Twitter in breaking events. For situations like election fraud, civil violence, and state terrorism, Twitter could be a very powerful tool.

A significant blow to activists in the developing world, however, has been the discontinuation of the sms service of Twitter to all countries other than the U.S., Canada, and India. This means that people cannot use the very simple and widespread interface—the common everyday texting service—to post to Twitter.  As Twitter grew into a worldwide phenomenon, the cost for sms service worldwide was simply too great for the Twitter company to bear [11]. Twitter is in communication with mobile companies in other nations to negotiate some cost sharing that would make this service commercially viable in places outside of the U.S., Canada and India, but until then Twitterers outside of those countries will have to use browser-enabled, email-enabled, iPhones or Blackberries to communicate with the Twit-o-sphere, if they are going to do any mobile tweeting. But all is not lost. Twitter competitors are emerging: companies: Jaiku offers sms service; a Twitter knock-off, it hopes to become the service of choice for those whom the sms feature was an important, or even critical feature. A trickle of migration is said to be occurring.

The potential for Twitter and blogging to affect the political landscape, particularly of places like Africa, where communication networks are scarce and precious resources, is beginning to emerge. The situation is one of what Yaneer Bar-Yam calls "distributing complexity". He observes:

*When we want to accomplish a task, the complexity of the system performing that task must match the complexity of the task. In order to perform the matching correctly, one must recognize that each person has a limited level of complexity. Therefore, tasks become difficult because the complexity of a person is not large enough to handle the complexity of the task. The trick then is to distribute the complexity of the task among many individuals. This is both similar and different from the old approach of distributing the effort involved in a task. When we need to perform a task involving more effort than one person can manage, for example, lifting a heavy object, then a group of people might all lift at the same time to distribute the effort [12].*

The old adage, "Many hands make light work" certainly applies, but in as complex and delicate a task as changing the entrenched power structures, confronting corruption and developing a new political voice, more than just sheer numbers of people is required. It is more important that the *right* people find one another. Or that a sufficient number and type of people are connected.

In Africa, particularly, there is a generational divide that further complicates the task. George Ayittey, a prominent economist, has divided the generations of Africa into "hippos" and "cheetahs." "Hippos" are the older generation, who place the blame for Africa's problems on the doorstep of Western colonialists; "cheetahs" –the younger generation—are more apt to see the problem in the predatory elites that run their national apparatuses. These young (and old) activists have in the past been isolated from one another. The threat of prison or exile is a real one [13].

The tools of social computing can provide the long distance support necessary to "lift the elephant" of the past with a relatively small number of people. No where is that more clean than in the December 2008 Ghanaian elections. A very close, hard fought campaign in a deeply divided country, election worries were significant. In a curious technological innovation, sms messaging was used to monitor 1000 polling stations by CODEO, the Coalition of Domestic Election Observers (http://www.codeogh.org/) with an election observation room in Accra at the Kofi Annan International Peacekeeping Training Center. Over 4000 trained, local observers, wearing white teeshirts with CODEO ELECTION OBSERVER- ELECTION 2008 emblazoned on their backs, descended on the 230 constituencies all over Ghana. As reported by MobileActiv.org, systematic sms-based reporting can provide "an independent and reliable indicator about the quality of the election process [14]. The elections were very successful, with sms messaging assisting in staunching rumors and providing a measure of confidence in the election returns.

Will technology enable the cheetahs?   Watching the latest news, the answer would seem to be, "yes."   Keeping up with the cheetahs is going to be hard for the technologically less gifted hippos—and despite the small number of people involved, technology may provide the needed "pulley" to move nation states swiftly in new directions. Twitter may pass away to the next big thing, but certainly, it is no understatement to say that  social computing technologies are changing the world.

# References

1.  Yaneer, Bar-Yam (2005), Solving Real World Problems. In Making Things Work: Solving Complex Problems: Solving Complex Problems in a Complex World.
2.  TinySex, the precursor of cybersex or "cyber".
3.  Beaumont, Claudine (2008), Mumbai attacks: Twitter and Flickr used to break news. Telegraph. http://www.telegraph.co.uk/news/worldnews/asia/india/3530640/Mumbai-attacks-Twitter-and-Flickr-used-to-break-news-Bombay-India.html
4.  Sarno, David (2008), Mumbai News Fished from Twitter's Rapids. LA Times http://latimesblogs.latimes.com/technology/2008/12/mumbai-news-fis.html
5.  Ingram, Matthew (2008), Yes, Twitter is a source of journalism. www.mathewingram.com/work/2008/11/26/yes-twitter-is-a-source-of-journalism/
6.  Gladkova, Svetlana (2008), Mumbai Attacks: Twitter Adds to the Noise but Is Still Valuable. http://profy.com/2008/11/27/mumbai-attacks-twitter-adds-to-the-noise-but-is-still-valuable/
7.  Ranganathan, Vinu. http://flickr.com/photos/vinu/sets/72157610144709049/
8.  Vandermore, Snow. 10 Twitter Agendas: what's yours? http://www.twitip.com/10-twitter-agendas-whats-yours/#more-707
9.  http://www.pbs.org/mediashift/2008/03/how-bloggers-covered-kenya-violence-deal-with-racism-sexism085.html
10. http://www.pbs.org/mediashift/2008/11/nigeria-joins-list-of-countries-harassing-bloggers315.html
11. Twitter (2008), Changes for SMS Users—Good and Bad News. http://blog.twitter.com/2008/08/changes-for-some-sms-usersgood-and-bad.html
12. Yaneer, Bar-Yam (2004), Solving World Problems. In: Making Things Work: Solving Complex Problems in a Complex World. NECSI Knowledge Press. P. 91.
13. Ayittey, George (2006), Africa Unchained: The Blueprint for Africa's Future
14. Verclas, Katrin (2008), SMS Critical to Election Monitoring in Ghana. http://mobileactive.org/sms-critical-election-observation-ghana

# Rule Evaluation Model as Behavioral Modeling of Domain Experts

Hidenao Abe[1] and Shusaku Tsumoto[2]

[1]abe@med.shimane-u.ac.jp, Shimane University, JAPAN
[2]tsumoto@computer.org, Shimane University, JAPAN

**Abstract** In this paper, we present an experiment to describe behavior of a human decision on the rule evaluation procedure, which is a post-processing procedure in data mining process, based on objective rule indices. The post-processing of mined results is one of the key factors for successful data mining process. However, the relationship between transitions of human criteria and the objective rule evaluation indices has never been clarified as behavioral viewpoints. By using a method based on objective rule evaluation indices to support the rule evaluation procedure, we have evaluated the accuracies of five representative learning algorithms to construct rule evaluation models of the actual data mining results from a chronic hepatitis data set. Further, we discuss the relationship between the transitions of the subjective criteria of a medical expert and the rule evaluation models.

## 1 Introduction

In recent years, enormous amounts of data relating to natural science, social science, and business have been stored on information systems. In order to extract valuable knowledge from such databases, data mining techniques have been well known as one of the useful method. Data mining scheme are combined methods of different types of technologies such as database technologies, statistical methods, and machine learning methods. In particular, if-then rules, which are obtained by rule induction algorithms, are considered to be one of the widely usable and readable outputs of data mining. However, for large datasets with hundreds of attributes, including noise, the process often results in several thousands of rules. From such a large rule set, it is difficult for human experts to obtain valuable knowledge, which is rarely included in the rule set.

To support such a rule selection, many studies have been performed using objective rule evaluation indices such as recall, precision, and other interestingness measurements [1, 2, 3] (Hereafter, we refer to these indices as "objective indices"). Further, it is difficult to estimate the subjective criterion of a medical expert using a single objective rule evaluation index; this is because his/her subjective criterion such as interestingness and importance for his/her purpose is influenced by the amount of his/her medical knowledge and/or the passage of time.

H. Liu et al. (eds.), *Social Computing and Behavioral Modeling*,
DOI: 10.1007/978-1-4419-0056-2_3, © Springer Science + Business Media, LLC 2009

8

With regard to the above mentioned issues, we have developed an adaptive rule evaluation support method for human experts with rule evaluation models. This method predicts the experts' criteria by using a model learned from the values of objective indices and the results of the evaluations given by human experts.

In Section 2, we describe the rule evaluation support method based on rule evaluation models. Then, we present a performance comparison of learning algorithms for combining rule evaluation models in Section 3. Then, we discuss the relationship between the transitions of each subjective criteria of a medical expert and the observed differences of the rule evaluation models.

## 2 Rule Evaluation Support with Rule Evaluation Models Based on Objective Indices

In practical data mining situations, costly rule evaluation procedures are repeatedly performed by a human expert. In these situations, the useful information from each evaluation, such as focused attributes, interesting combinations of attributes/attributes and values, and valuable facts are not explicitly used by any rule selection system, but tacitly assumed by the human expert. To solve these problems, we suggest a method to combine rule evaluation models based on objective rule evaluation indices as a way to explicitly describe the criteria of the human expert by re-using the human evaluations. By combining this method with the rule visualization interface, we design a rule evaluation support tool, which can perform more certain rule evaluations with explicit rule evaluation models.

We consider the process of modeling the rule evaluations of human experts as the process that clarifies the relationships between the human evaluations and the features of the inputted if-then rules. Based on this consideration, we find that the construction process for the rule evaluation model can be implemented as a learning task. Figure 1 shows the construction process based on the re-use of human evaluations and objective indices for each mined if-then rule.

Using this framework, the human expert can evaluate a large number of if-then rules iteratively. The system supports his/her rule evaluation by presenting predictions for unevaluated rules based on a rule evaluation model learned from the evaluated rules. Thus, the human expert can avoid the evaluation of unfamiliar rules or/and uncertain rules based on his/her background knowledge from the first time evaluation process of each rule.

In the training phase, the attributes of a meta-level training dataset are obtained by objective indices such as recall, precision, and other rule evaluation values. The human evaluations for each rule are combined as classes of each instance. To obtain this data set, the human expert has to evaluate the all or a part of the input rules at least once. After obtaining the training data set, its rule evaluation model is constructed by using a learning algorithm.

In the prediction phase, the human expert receives the predictions for new rules based on their objective index values. Since the rule evaluation models are used for

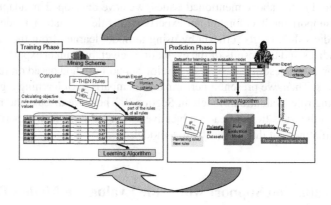

**Fig. 1** Training phase and prediction phase of the rule evaluation support method.

predictions, we need to choose a learning algorithm with high accuracy in order to solve the current classification problems.

## 3 A Case Study on Chronic Hepatitis Data Mining Results

In this section, we present the results of empirical evaluations using the dataset obtained from two hepatitis data mining results [4, 5]. First, we obtained a dataset for each rule set from the above-mentioned data mining results. The five learning algorithms were then used to combine each rule evaluation model.

Based on the experimental results, we evaluated the availability of each learning algorithm with respect to the accuracies of the rule evaluation models and the contents of the rule evaluation models learned.

Since a rule evaluation model requires high accuracy to perfectly support a human expert in the proposed rule evaluation support method, we compared the predictive accuracies for the entire dataset and Leave-One-Out validation.

We obtained the learning curves of accuracies of the learning algorithms for the entire training dataset to evaluate whether each learning algorithm can perform in the early stage of the rule evaluation process.

By observing the elements of the rule evaluation models for the meningitis and hepatitis data mining results, we consider the characteristics of the objective indices that are used in these rule evaluation models.

In order to obtain a dataset to learn a rule evaluation model, the values of the objective indices have been calculated for each rule by considering 39 objective indices [7], as shown in Table 1. Thus, the dataset for each rule set has the same

**Table 1** Objective rule evaluation indices for classification rules used in this research. **P:** Probability of the antecedent and/or consequent of a rule. **S:** Statistical variable based on P. **I:** Information of the antecedent and/or consequent of a rule. **N:** Number of instances included in the antecedent and/or consequent of a rule. **D:** Distance of a rule from the others based on rule attributes.

| Theory | Index Name (**Abbreviation**) [Reference Number of Literature] |
|---|---|
| P | Coverage (**Coverage**), Prevalence (**Prevalence**) |
| | Precision (**Precision**), Recall (**Recall**) |
| | Support (**Support**), Specificity (**Specificity**) |
| | Accuracy (**Accuracy**), Lift (**Lift**) |
| | Leverage (**Leverage**), Added Value (**Added Value**) |
| | Klösgen's Interestingness (**KI**), Relative Risk (**RR**) |
| | Brin's Interest (**BI**), Brin's Conviction (**BC**) |
| | Certainty Factor (**CF**), Jaccard Coefficient (**Jaccard**) |
| | F-Measure (**F-M**), Odds Ratio (**OR**) |
| | Yule's Q (**YuleQ**), Yule's Y (**YuleY**) |
| | Kappa (**Kappa**), Collective Strength (**CST**) |
| | Gray and Orlowska's Interestingness (**GOI**) |
| | Gini Gain (**Gini**), Credibility (**Credibility**) |
| S | $\chi^2$ Measure for One Quadrant ($\chi^2$-**M1**) |
| | $\chi^2$ Measure for Four Quadrants ($\chi^2$-**M4**) |
| I | J-Measure (**J-M**), K-Measure (**K-M**)[6] |
| | Mutual Information (**MI**) |
| | Yao and Liu's Interestingness 1 based on one-way support (**YLI1**) |
| | Yao and Liu's Interestingness 2 based on two-way support (**YLI2**) |
| | Yao and Zhong's Interestingness (**YZI**) |
| N | Cosine Similarity (**CSI**), Laplace Correction (**LC**) |
| | $\phi$ Coefficient ($\phi$) |
| | Piatetsky-Shapiro's Interestingness (**PSI**) |
| D | Gago and Bento's Interestingness (**GBI**) |
| | Peculiarity (**Peculiarity**) |

number of instances as the rule set. Each instance has 40 attributes, including those of the class.

We applied the five learning algorithms to these datasets to compare their performances as construction methods for the rule evaluation models. We employed the following learning algorithms from Weka [8]: J4.8 (a C4.5 decision tree learner) [9], a back-propagation neural network (BPNN) learner [10], support vector machines (SVM) [11], classification via linear regressions (CLR) [12], and OneR [13].

## 3.1 Description of the Chronic Hepatitis Data Mining Results

In this case study, we used four data mining results for chronic hepatitis, as shown in Table 2. These datasets show patterns of the values obtained from the laboratory tests of the blood and urine of chronic hepatitis patients as attributes.

First, we performed the data mining processes twice to determine the relationships between the patterns of attributes and the those of glutamate-pyruvate transaminase (GPT) as the class, which is one of the important tests required to determine the condition of each patient. A medical expert evaluated these results at different times, which were approximately one year apart. However, the expert held his interestingness concerning the significant movement of GPT values between these two evaluations. Second, we performed other data mining processes twice to determine the relationships between the patterns of attributes and the results of interferon (IFN) therapy. The expert also evaluated these results at different times, which were two weeks apart. For each rule, we assigned a label (EI: Especially Interesting, I: Interesting, NI: Not Interesting, and NU: Not Understandable) according to the evaluations provided by the medical expert.

**Table 2** Description of datasets of the chronic hepatitis data mining results.

| | | #Rules | Class Distribution | | | |
|---|---|---|---|---|---|---|
| | | | EI | I | NI | NU |
| G P T I F N | Phase_1 | 30 | 3 | 8 | 16 | 3 |
| | Phase_2 | 21 | 2 | 6 | 12 | 1 |
| | First Time | 26 | 4 | 7 | 11 | 4 |
| | Second | 32 | 15 | 5 | 11 | 1 |

## 3.2 Comparison of the classification performances

The results for the performances of the five learning algorithms with respect to the entire training dataset and those for Leave-One-Out are shown in Tables 3, respectively. Most of the accuracies for the entire training dataset are higher than those estimated by simply predicting each default class. This indicates that the five learning algorithms can construct valid rule evaluation models for these datasets.

**Table 3** Accuracies (%), Recalls (%) and Precisions (%) of the five learning algorithms for the training datasets.

| | | Entire Dataset | | | | | | | | | Leave-One-Out | | | | | | | |
|---|---|---|---|---|---|---|---|---|---|---|---|---|---|---|---|---|---|---|
| | Acc | Precision | | | | Recall | | | | Acc | Precision | | | | Recall | | | |
| | | EI | I | NI | NU | EI | I | NI | NU | | EI | I | NI | NU | EI | I | NI | NU |
| **GPT1** | | | | | | | | | | | | | | | | | | |
| J4.8 | 96.7 | 100.0 | 88.9 | 100.0 | 100.0 | 66.7 | 100.0 | 100.0 | 100.0 | 50.0 | 0.0 | 60.0 | 60.0 | 0.0 | 0.0 | 75.0 | 56.3 | 0.0 |
| BPNN | 100.0 | 100.0 | 100.0 | 100.0 | 100.0 | 100.0 | 100.0 | 100.0 | 100.0 | 30.0 | 0.0 | 12.5 | 50.0 | 0.0 | 0.0 | 12.5 | 50.0 | 0.0 |
| SVM | 56.7 | 0.0 | 100.0 | 68.2 | 14.3 | 0.0 | 12.5 | 93.8 | 33.3 | 46.7 | 0.0 | 0.0 | 65.0 | 11.1 | 0.0 | 0.0 | 81.3 | 33.3 |
| CLR | 63.3 | 0.0 | 66.7 | 62.5 | 0.0 | 0.0 | 50.0 | 93.8 | 0.0 | 40.0 | 0.0 | 14.3 | 50.0 | 0.0 | 0.0 | 12.5 | 68.8 | 0.0 |
| OneR | 60.0 | 0.0 | 66.7 | 59.3 | 0.0 | 0.0 | 25.0 | 100.0 | 0.0 | 43.3 | 0.0 | 25.0 | 55.6 | 0.0 | 0.0 | 37.5 | 62.5 | 0.0 |
| **GPT2** | | | | | | | | | | | | | | | | | | |
| J4.8 | 90.5 | 66.7 | 85.7 | 100.0 | 0.0 | 100.0 | 100.0 | 91.7 | 0.0 | 76.2 | 0.0 | 66.7 | 90.9 | 0.0 | 0.0 | 100.0 | 83.3 | 0.0 |
| BPNN | 100.0 | 100.0 | 100.0 | 100.0 | 100.0 | 100.0 | 100.0 | 100.0 | 100.0 | 66.7 | 0.0 | 83.3 | 81.8 | 0.0 | 0.0 | 83.3 | 75.0 | 0.0 |
| SVM | 95.2 | 100.0 | 100.0 | 92.3 | 100.0 | 50.0 | 100.0 | 100.0 | 0.0 | 81.0 | 0.0 | 100.0 | 91.7 | 25.0 | 0.0 | 83.3 | 91.7 | 100.0 |
| CLR | 85.7 | 50.0 | 100.0 | 85.7 | 0.0 | 50.0 | 83.3 | 100.0 | 0.0 | 76.2 | 0.0 | 83.3 | 84.6 | 0.0 | 0.0 | 83.3 | 91.7 | 0.0 |
| OneR | 85.7 | 0.0 | 75.0 | 92.3 | 0.0 | 0.0 | 100.0 | 100.0 | 0.0 | 81.0 | 0.0 | 66.7 | 91.7 | 0.0 | 0.0 | 100.0 | 91.7 | 0.0 |
| **INF1** | | | | | | | | | | | | | | | | | | |
| J4.8 | 88.5 | 80.0 | 100.0 | 83.3 | 100.0 | 100.0 | 71.4 | 90.9 | 100.0 | 19.2 | 37.5 | 0.0 | 20.0 | 0.0 | 75.0 | 0.0 | 18.2 | 0.0 |
| BPNN | 100.0 | 100.0 | 100.0 | 100.0 | 100.0 | 100.0 | 100.0 | 100.0 | 100.0 | 26.9 | 40.0 | 22.2 | 25.0 | 25.0 | 50.0 | 28.6 | 18.2 | 25.0 |
| SVM | 46.2 | 26.7 | 0.0 | 70.0 | 100.0 | 100.0 | 0.0 | 63.6 | 25.0 | 34.6 | 21.4 | 0.0 | 54.5 | 0.0 | 75.0 | 0.0 | 54.5 | 0.0 |
| CLR | 53.8 | 100.0 | 0.0 | 47.6 | 66.7 | 50.0 | 0.0 | 90.9 | 50.0 | 19.2 | 33.3 | 0.0 | 28.6 | 0.0 | 25.0 | 0.0 | 36.4 | 0.0 |
| OneR | 50.0 | 0.0 | 50.0 | 50.0 | 0.0 | 0.0 | 85.7 | 63.6 | 0.0 | 19.2 | 0.0 | 11.1 | 23.5 | 0.0 | 0.0 | 14.3 | 36.4 | 0.0 |
| **INF2** | | | | | | | | | | | | | | | | | | |
| J4.8 | 90.6 | 88.2 | 100.0 | 90.9 | 0.0 | 100.0 | 80.0 | 90.9 | 0.0 | 75.0 | 76.5 | 66.7 | 75.0 | 0.0 | 86.7 | 40.0 | 81.8 | 0.0 |
| BPNN | 100.0 | 100.0 | 100.0 | 100.0 | 100.0 | 100.0 | 100.0 | 100.0 | 100.0 | 37.5 | 50.0 | 28.6 | 22.2 | 0.0 | 53.3 | 40.0 | 18.2 | 0.0 |
| SVM | 56.3 | 72.7 | 0.0 | 45.0 | 100.0 | 53.3 | 0.0 | 81.8 | 100.0 | 31.3 | 36.4 | 0.0 | 28.6 | 0.0 | 26.7 | 0.0 | 54.5 | 0.0 |
| CLR | 65.6 | 63.2 | 100.0 | 60.0 | 0.0 | 80.0 | 60.0 | 54.5 | 0.0 | 34.4 | 41.2 | 20.0 | 30.0 | 0.0 | 46.7 | 20.0 | 27.3 | 0.0 |
| OneR | 68.8 | 62.5 | 0.0 | 87.5 | 0.0 | 100.0 | 0.0 | 63.6 | 0.0 | 68.8 | 60.0 | 0.0 | 100.0 | 0.0 | 100.0 | 0.0 | 63.6 | 0.0 |

Besides, the accuracies of Leave-One-Out show the robustness of each learning algorithm. These accuracies show the predictive performances of each learning algorithm in the "Prediction Phase" in Figure 1. For GPT1 and GPT2, the medical expert is required to determine "EI" rules by himself/herself, because the learning algorithms cannot predict "EI" instances based on the training datasets. Besides, for GPT1 and IFN1, the accuracies are lower than those simply estimated by predicting default classes because the medical expert evaluates these data mining results without a specific criterion in his/her mind.

## 3.3 Rule evaluation models for the data mining result datasets of chronic hepatitis

Figure 2 shows the decision trees of J4.8 for each data mining result dataset for chronic hepatitis. With a decision tree, the class label (label of squared node) of an instance is predicted based on its attribute values of the instance according to each branch of the tree. Hence, these decision trees show the decision models of the medical expert from one of the objective aspects. As shown in Figure 2 and Figure 3, these models consist of not only indices expressing the correctness of rules but also other types of indices. This shows that the medical expert evaluated these rules with both correctness and interestingness based on his/her background knowledge. In each problem, the variance of indices was reduced in the second data mining process as shown in Figure3 .Figure 3 shows the statistics of the frequencies of each indices, which appeared in the decision trees in 10,000 times bootstrap iterations. This indicates that the medical expert evaluated the second data mining results with a more certain criterion than he did for the first data mining process.

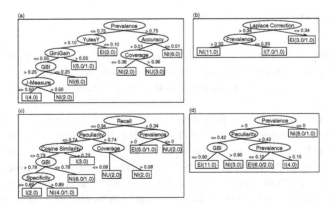

**Fig. 2** Rule evaluation models constructed using J4.8 for each data mining result ( (a)GPT phase 1, (b) GPT phase 2, (c) IFN first time, (d) IFN second time).

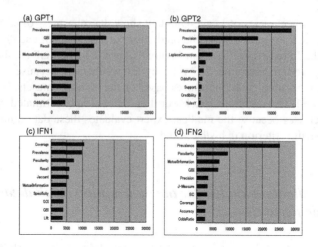

**Fig. 3** Frequencies of objective indices appeared in 10,000 time bootstrap re-sampling iterations of J4.8 decision trees ( (a)GPT phase 1, (b) GPT phase 2, (c) IFN first time, (d) IFN second time).

The difference between the hepatitis data mining results concerns the certainness of the criterion of the medical expert. The medical expert built up his/her criterion by using iterating data mining processes. Based on these differences, we identified the following three phases of rule evaluation: hypothesis generation, hypothesis validation, and hypothesis refinement.

Then, by considering the differences in performances and the contents of the rule evaluation models, we defined the relationship between these observable differences and the three evaluation phases.


# 4 Conclusion

In the present paper, we have described the evaluation of five learning algorithms for a rule evaluation support method; rule evaluation models based on objective indices are used to predict the evaluations for an if-then rule by re-using the evaluations of a human expert.

Based on the comparison of performances of the five learning algorithms for the dataset from the data mining results of chronic hepatitis, the rule evaluation models have achieved higher accuracies for simply predicting each default class. Considering the results of the rule evaluation models constructed for four different data mining results of chronic hepatitis, the differences in the human evaluation criteria appear as differences in the rule evaluation models with respect to both their performances and their contents. This result indicates that the proposed approach can detect differences in the human evaluation criteria as differences in the performances of rule evaluation models.

In the future, we will apply this rule evaluation support method to other datasets from various domains.

# References

1. Hilderman R.J. and Hamilton H.J. (2001) Knowledge discovery and measures of interest. Kluwer Academic Publishers
2. Tan P.N., Kumar V., and Srivastava J. (2003) Selecting the right interestingness measure for association patterns. In: Proc. of International Conference on Knowledge Discovery and Data Mining, 32–41
3. Yao Y.Y. and Zhong N. (1999) An analysis of quantitative measures associated with rules. In: Proc. of Pacific-Asia Conference on Knowledge Discovery and Data Mining, PAKDD-1999, 479–488
4. Ohsaki M., Sato Y., Kume S., Yokoi H., and Yamaguchi T. (2004) Comparison between objective interestingness measures and real human interests in medical data mining. In: Proc. of the 17th International Conference on Industrial and Engineering Applications of Artificial Intelligence and Expert Systems IEA/AIE-2004, LNAI 3029, 1072–1081
5. Abe H., Ohsaki M., Yokoi H., and Yamaguchi T. (2006) Implementing an integrated time-series data mining environment based on temporal pattern extraction methods?A case study of an interferon therapy risk mining for chronic hepatitis, In: JSAI2005 Workshops, LNAI 4012, 425–435
6. Ohsaki M., Kitaguchi S., Kume S., Yokoi H., and Yamaguchi T. Evaluation of rule interestingness measures with a clinical dataset on hepatitis. In: Proc. of ECML/PKDD-2004, LNAI 3202, 362–373
7. Ohsaki M., Abe H., Yokoi H., Tsumoto S., and Yamaguchi T. (2007) Evaluation of Interestingness Measures in Medical Knowledge Discovery in Databases, Artificial Intelligence in Medicine, 41(3), 177–196
8. Witten I.H. and Frank E. (2000) Data Mining: Practical machine learning tools and techniques with Java implementations, Morgan Kaufmann
9. Quinlan J.R. (1993) Programs for machine learning, Morgan Kaufmann
10. Runmelhart D. E., McClelland J. L. (1986) Parallel Distribute Processing, MIT Press
11. Platt J. (1999) Fast training of support vector machines using sequential minimal optimization. In: Schoelkopf B., Burges C., and Smola A. (eds.), Advances in Kernel Methods-Support Vector Learning, MIT Press, 185–208
12. Frank, E., Wang, Y., Inglis, S., Holmes, G., and Witten, I.H. (1998) Using model trees for classification. Machine Learning 32(1) 63–76
13. Holte R.C. (1993) Very simple classification rules perform well on most commonly used datasets. Machine Learning, 11, 63?-91

# Trust and Privacy in Distributed Work Groups

Denise Anthony[1], Tristan Henderson[2], and James Kitts[3]

[1] denise.anthony@dartmouth.edu, Dartmouth College, Hanover, NH, USA
[2] tristan@cs.st-andrews.ac.uk, University of St. Andrews, St. Andrews, Scotland, UK
[3] jak2190@columbia.edu, Columbia University, New York, NY, USA

**Abstract** Trust plays an important role in both group cooperation and economic exchange. As new technologies emerge for communication and exchange, established mechanisms of trust are disrupted or distorted, which can lead to the breakdown of cooperation or to increasing fraud in exchange. This paper examines whether and how personal privacy information about members of distributed work groups influences individuals' cooperation and privacy behavior in the group. Specifically, we examine whether people use others' privacy settings as signals of trustworthiness that affect group cooperation. In addition, we examine how individual privacy preferences relate to trustworthy behavior. Understanding how people interact with others in online settings, in particular when they have limited information, has important implications for geographically distributed groups enabled through new information technologies. In addition, understanding how people might use information gleaned from technology usage, such as personal privacy settings, particularly in the absence of other information, has implications for understanding many potential situations that arise in pervasively networked environments.

## 1 Introduction

Trust plays an important role in group cooperation. During periods of broad social change, however, the basis of trust, and therefore the ability for social actors to engage in exchange and cooperation, can be disrupted. For example, during the Industrial Revolution, increased contact and interaction between unknown individuals as a result of immigration to cities amplified uncertainty regarding the reliability and trustworthiness of potential exchange partners.[1] Similarly, interaction occurred in new settings and situations, e.g., factories, in which individuals' behavior and outcomes depended on the actions of possibly unknown others.[1,2] Over time, new mechanisms were created to detect, monitor and signal the reliability/trustworthiness of social actors.[1,2] Today, as new information technologies (IT) emerge for communication and exchange that facilitate contact between unknown individuals in novel settings and situations (e.g., chat rooms, social networking websites, distributed work groups), established mechanisms of trust are disrupted or distorted, which can lead to the breakdown of cooperation or to increasing fraud in exchange. Moreover, if

H. Liu et al. (eds.), *Social Computing and Behavioral Modeling*,
DOI: 10.1007/978-1-4419-0056-2_4, © Springer Science + Business Media, LLC 2009

new mechanisms for determining or signaling trustworthiness have yet to be established, individuals may use other signals as a basis for trust; signals which may or may not be associated with reliability yet will affect interaction, exchange and cooperation.

Interest in the implications of information technology for trust crosses all of the social science disciplines.[3-11] Social scientists, for example, have explored how interpersonal trust is adapted to the digital environment[6], e.g., the reputation system of *eBay* [7] and reliability in the online encyclopedia *Wikipedia.org* [12]; how trust affects consumer and other behavior online [8,13-15]; and how institutional trust is managed in pervasive networks by organizations such as firms. [9,16-17] Here, we build on these studies to examine how personal privacy information about members of geographically distributed, virtual work groups influences cooperation and privacy behavior in the group. Specifically, we use experimental methods to examine whether people use others' personal privacy settings as signals of trustworthiness within the group that affect cooperative group behavior. In addition, we examine how individual privacy preferences relate to whether individuals cooperate in the group. Understanding how people interact with others in online settings, in particular when they have limited information, has important implications for geographically distributed groups enabled through new information technologies. In addition, understanding how people might use information gleaned from technology usage, such as personal privacy settings, particularly in the absence of other information, has implications for understanding many potential situations that arise in pervasively networked environments.

In section 2 we briefly discuss the social science literature on trust, as well as previous research on trust in work groups, including distributed teams facilitated with IT support. Section 3 describes the social experiment used for the study and the characteristics of the subjects. Section 4 presents results, while section 5 discusses the findings, limitations and implications.

## 2 Trust, Cooperation and Work Teams

Trust is a term to describe positive expectations of one actor toward another for some specific action. When we say that A trusts B, we typically mean that A trusts B to do X [18-19], and so A may take some action Y (e.g., lending $10, sharing information, contributing to a joint project) in which A is now vulnerable to losing Y depending on the behavior of B (i.e., doing X or not). According to Edward Lorenz [20], A's behavior Y based on trust in B "consists in action that (1) increases one's vulnerability to another whose behavior is not under one's control, and (2) takes place in a situation where the penalty suffered if the trust is abused would lead one to regret the action." Snijders[21] specifies that an actor's vulnerability in trust relationships is based on uncertainty about another actor's "disposition or preferences" for cooperation, not his or her abilities to cooperate. [2, 22-23]

A's positive expectations about B, i.e., A's perception of B's trustworthiness, may be based on a number of different reasons, including: (1) A's past experience with B in general or on X specifically; (2) A's relationship with B; (3) A's knowledge of B's reputation from other actors; (4) A's knowledge of B's incentives to do X in response to some other third-party. However, social actors may have very limited information about potential exchange partners (B) and therefore have limited grounds on which to assess trustworthiness. Indeed, as noted above, during periods of rapid social change, social actors may have much greater contact with unknown others for whom they have limited information. In such cases of limited information, some characteristics or behavior may be perceived by others as signals of trustworthiness, regardless of the association between such characteristics and behavior. For example, Zucker [1] describes how some actors may perceive characteristics such as race or gender as signals of trustworthiness, without the knowledge or intention of the so-characterized actor.

       Trust and work teams. IT and work teams
       Trust and IT.

# 3 Methods

In the summer of 2007 we conducted a social experiment in which undergraduate subjects (n=110) participated in a series of simulated online groups. After responding to a recruitment advertisements and completing an IRB-approved Informed Consent form, undergraduate student subjects completed a brief online pre-survey measuring demographic characteristics, experience in Internet commerce, and attitudes toward taking risks. Upon completing the pre-survey, each subject was assigned a numeric identifier used to anonymously link the pre-survey responses to the experiment results and post-survey responses. Subjects brought their identification number to the lab session where they participated in the experiment session that lasted approximately 20 minutes.

       In the lab session, subjects were told they were members of geographically distributed work teams (with two other individuals not known by the subjects) that were working on developing a proprietary product via a secure (password protected) online project *wiki*. In addition, the team was in competition with other project teams for the best product design, so project development information was even more closely guarded than standard proprietary designs. Given that the project *wikis* contained valuable information, team members were faced with opportunities to sell the password. Subjects made decisions about whether to sell the *wiki* password in a series of six different teams, comprised of different types of group members (described further below), with each team engaged in a competition as described. Subjects earned points based on the outcomes of each team-competition (other group members are simulated so their behavior is determined randomly); points were converted to dollar amounts that subjects were paid upon completing the study. If any team member sold the password, all team members received zero points. If the subject sold the password, s/he was awarded points according to one of two randomly assigned conditions: (1) 4 points, which was less than the potential gain from group success while protecting the password (6 points) or winning the competition (8 points); or (2) 8 points, which was greater than the potential gain from group success while protecting password (6 points) and equal to points from winning the competition (8 points). If the subject chose not to sell the password, s/he was awarded either 0, 6 or 8 points depending on, respectively, if another group member sold the password (20% chance, randomly determined), if no member sold the password (70% chance, randomly determined), or if no member sold the password *and* the team won the competition (10% chance, randomly determined).

For each team, subjects receive two pieces of information about the other group members: (1) individual's privacy setting on a scale of 1 – 3 (1=Private 2=Moderate and 3=Open), and (2) individual's skill level: Low or High. Subjects were teamed with two (simulated) members and decisions about whether to sell the password were made simultaneously with no interaction allowed. In order to reduce the chance that feedback about outcomes would influence subjects' subsequent decisions, subjects were not told the number of teams they would participate in, or the outcomes for their teams until the end of the experiment.

# 3 Results

A total of 110 subjects participated in the study, with each subject making choices in 6 rounds (different team configurations) to decide whether to sell the password or not (n=110*6 = 660 observations).

Our central research question is whether subjects use others' privacy settings as a signal of trustworthiness, in particular when they have limited information about group members. If so, we would expect to see that subjects behave differently depending on group members' privacy settings. In the experimental setting described here, we tested whether subjects were more or less likely to cooperate (*not* sell the group password) depending on other group members' privacy settings of open, moderate or private, and controlling for other factors, including subjects' own privacy setting (explored further below), group members' skill level, and value of group cooperation (i.e., size of the incentive to sell based on whether value of selling password $\geq$ potential group payoff).

Table 1 shows the percentage of subjects who were willing to sell the password (*not* cooperate) depending on the privacy settings of teammates, after controlling for the effect of subjects' own privacy setting, teammates' skill level, and the value of group cooperation. Subjects were significantly more likely to sell the password when paired with a teammate who was more open, and even more likely to sell when paired with two teammates who were more open. (P-values based on estimates from logistic regression of choice to sell password on teammates' privacy settings, controlling for subject's privacy setting, teammates' skill levels, and size of the incentive to sell, with robust standard errors.) That is, subjects were most likely to sell the password when paired with two open teammates, and least likely to sell the password when paired with two private teammates. These findings suggest that subjects do indeed use other's privacy settings as signals of trustworthiness in conditions of limited information, altering their behavior in ways that indicate they believe others who have more open settings are less trustworthy.

**Table 1.** Percent willing to sell password by team members' privacy settings

| Teammate 2: Privacy Setting | Teammate 1: Privacy Setting | | | P-value |
|---|---|---|---|---|
| | OPEN | MODERATE | PRIVATE | |
| OPEN | 58% | 45% | --- | Teammate 1 P<.001 |
| MODERATE | --- | 39% | --- | Teammate 2 P<.001 |
| PRIVATE | 46% | 34% | 24% | |

*Note:* P-values based on estimates from logistic regression of choice to sell password on teammates' privacy settings, controlling for subject's privacy setting, teammates' skill levels, and size of the incentive to sell, with robust standard errors.

These findings raise the question as to whether individuals who are more open are indeed less trustworthy, i.e., are subjects with open privacy settings more likely to sell the group password than those with more private settings? Table 2 shows the mean differences (unadjusted) in likelihood of selling the password between subjects with different privacy settings. In contrast to the findings above, subjects who are more private are somewhat *more* likely to sell the password (p<.10) than subjects who are open.

**Table 2.** Percent willing to sell password by subject's privacy setting

| Subject's Privacy Setting | % willing to sell password | ANOVA |
|---|---|---|
| Private | 48 | F = 2.56 P<.10 |
| Moderate | 40 | |
| Open | 36 | Bonferroni Private > Open P<.10 |

We examine this finding further in Figure 1, which shows the interaction between subject's privacy setting (comparing only Open with Private, suppressing Moderate category) and Teammates' privacy settings (comparing only teams in which teammates are either both Open or both Private). Consistent with the findings in Table 1, Figure 1 shows that all subjects are less likely to cooperate with open teammates than with private teammates. Surprisingly, however, when faced with two private teammates, private subjects are less likely to cooperate than open subjects. These preliminary findings suggest that though all subjects appear to cooperate more with private teammates, private subjects are least likely to actually cooperate!

**Figure 1.** Percent willing to sell password by interaction of Subject privacy setting with teammates' privacy settings

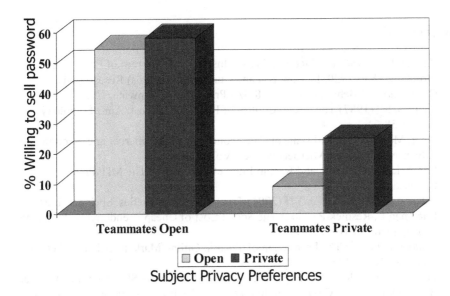

## 5 Discussion and Conclusion

Others' privacy preferences affect trust; those with more private settings are more likely to be viewed as trustworthy, and therefore to be trusted. Users' own privacy preferences also appear to matter for trust; more private users appear to be less trustworthy with regard to protecting group privacy and/or to be more distrustful of others than are users with more open settings.

Users will use privacy settings as "signals" of trustworthy behavior in groups, but those signals are not necessarily accurately associated with trustworthy behavior. Managing privacy in online groups or the "commons" enabled by new information technologies may be more difficult than expected as it does not appear to be a simple aggregate of individual preferences or behavior. In short, social dynamics and social context are likely to matter as much (or more) for ensuring privacy and security than technology alone.

**Acknowledgements** This research was supported by the Institute for Security Technology Studies at Dartmouth College under grant 2005-DD-BX-1091 awarded by the US Bureau of Justice Assistance. Points of view or opinions in this document are those of the authors and do not represent the official position or policies of the US Department of Justice or of other sponsors. We thank Clare Fortune-Agan, Linda Lomelino and Sara del Nido for research assistance.

# References

1. Zucker, LG (1986) Production of Trust: Institutional Sources of Economic Structure, 1840-1920. In Staw BM and Cummings LL (eds) Research in Organizational Behavior, volume 8. JAI Press Inc., Greenwich, CT.
2. Shapiro, SP (1987) The Social Control of Impersonal Trust. Am J Sociology 93:623-58.
3. Baye, M (2002) Special Issue on The economics of the Internet and e-commerce. Advances in Applied Microeconomics. Volume 11.
4. Camp, LJ (2000) Trust and Risk in Internet Commerce. The MIT Press, Cambridge, MA.
5. Falcone R, Singh M, Tan YH (2001) Trust in Cyber-societies. Springer, Berlin.
6. Friedman E, Resnick P (2001) The Social Cost of Cheap Pseudonyms. J Econ & Management Str 10:173-199.
7. Kollock P, (1999) The Production of Trust in Online Markets. Advance in Group Processes 16:99-123.
8. Lunn R, Suman M (2002)Experience and Trust in Online Shopping. In: Wellman B, Haythornthwaite C (eds) The Internet in Everyday Life. Blackwell, Oxford UK.
9. Osterwalder D, (2001) Trust through evaluation and certification? Soc Sc Comp Rev 19:32-46.
10. Sambamurthy V, Jarvenpaa S, (eds) (2002). Special Issue on 'Trust in the Digital Economy.' J Strategic Inf Sys 11: 183-346.
11. Shapiro C, Varian H, (1999). Information Rules. A Strategic Guide to the Network Economy. Harvard Business School Press, Boston.
12. Anthony D, Smith SW, , Williamson T, (Forthcoming) Reputation and Reliability in Collective Goods: The case of the online encyclopedia Wikipedia. Rationality & Society.
13. Belanger F, Hiller J, Smith W, (2002) Trustworthiness in electronic commerce: the role of privacy, security and site attributes. J Strategic Info Sys 11: 245-270.
14. Castells M, (2001) The Internet Galaxy. Oxford University Press, Oxford.
15. Wellman B, Haythornthwaite C, (eds) (2002) The Internet in Everyday Life. Blackwell, Oxford.
16. Anthony D, Lewis E, Who do you call in a crisis? Reliability and Capability for Trusted Communication. Conference paper at the International Sunbelt Social Networks Conference XXIV, May 2004, Slovenia
17. Knights D, Noble F, Vurdubakis T, Willmott H, (2001) Chasing Shadows: Control, Virtuality and the Production of Trust. Org Studies 22:311-336

18. Hardin R (1991)
19. Hardin R (2001)
20. Lorenz E (1988)
21. Snijders C (1996) Trust and Commitments. Interuniversity Center for Social Science Theory and Methodology, Gronigen, Netherlands.
22. Luhmann N (1988) Familiarity, Confidence, Trust: Problems and Alternatives. In: Gambetta D (ed) Trust Making and Breaking Cooperative Relations. Basil Blackwell, New York.
23. Gambetta D (1988) Can We Trust Trust? In: Gambetta D (ed) Trust Making and Breaking Cooperative Relations. Basil Blackwell, New York.

# Cultural Consensus Theory: Aggregating Expert Judgments about Ties in a Social Network

William H. Batchelder[†]

† whbatche@uci.edu, Department of Cognitive Sciences, University of California Irvine

**Abstract** This paper describes an approach to information aggregation called Cultural Consensus Theory (CCT). CCT is a statistical modeling approach to pooling information from informants (experts, automated sources) who share a common culture or knowledge base. Each informant responds to the same set of questions, and the goal is to estimate the consensus knowledge of the informants. CCT has become a leading methodology for determining consensus beliefs of groups in the social sciences, especially cultural and medical anthropology. The paper illustrates CCT by providing a model for aggregating expert judgments about ties in a social network. Expert sources each provide a digraph on the same set of nodes, and the CCT model is used to estimate the most likely digraph to represent their shared knowledge.

## 1 Cultural Consensus Theory

Cultural Consensus Theory (CCT) is a statistical modeling approach to information aggregation developed by the author, A. Kimball Romney, and colleagues. It originated in the mid 1980s by a series of papers, e.g. [3], [4], [5], [16], [18]. Since its inception, CCT has become the leading methodology for determining consensus beliefs within the areas of social, cultural and medical anthropology; and it has also been applied in the areas of social networks, cognitive psychology, and psychometrics, e.g. [2], [9], [14], [15], [21]. In fact "Culture as consensus: A theory of culture and informant accuracy" [18], which first presented CCT to anthropological researchers, is the most highly cited article to appear in *American Anthropologist* in all the years since its origin (cf. Web of Science).

CCT consists of a collection of cognitively motivated, parametric statistical models for aggregating the responses of "informants" (experts, eye witnesses, automated sources) to a series of test items about some domain of their shared knowledge. It is assumed that there is a consensus (culturally correct) answer to each item based on the common culture (experiences) of the informants; however, these answers are not known to the researcher apriori and so they are specified as parameters in the models. Informants are not assumed to have complete access to the common knowledge, so the goal is to estimate that knowledge by pooling their responses. The shared knowledge could arise from a common language, common beliefs, common education, shared eye witness observations, shared religion, shared politics, and the like.

CCT works with questions in various formats where the answers represent the consensus views of a group rather than the idiosyncratic preferences of an individual. For example: "Do Hoosiers like corn?", "Is sinus contagious?", "Does Mr. A take orders from Mr. B?", or "What is the probability oil will be found in cite X?" are all reasonable items for a CCT analysis. However, "Do you like corn?", "Do you have sinus?", "Should Mr. A take orders from Mr. B?", or "Do you want to drill in site X?" are not appropriate items for CCT. Among the many actual and potential application areas for CCT are: 1. determining the shared beliefs of a deviant group; 2. determining social ties in a covert social network; 3. determining the syntax of an exotic language; 4. aggregating probability estimates from different sources; 5. cross-cultural comparisons of folk medical beliefs; 6. pooling eyewitness reports to a traumatic event; 7. detecting outliers in judges of sports competitions; 8. predicting the relative success of outreach efforts to shape opinion; and 9. measuring the cultural competence of experts separately from their idiosyncratic response biases.

CCT is intended to operate with relatively small numbers of heterogeneous informants. It applies in situations where researchers know how to pose questions in some format to the informants but they do not know the 'consensus answers' to the questions. The data structure of CCT requires that each informant answers each of a fixed series of questions. The informants may not be available for extensive training, do not experimentally interact prior to evaluation, and may be completely ignorant of English or, more generally, of any technical information (as required, say, in elicited priors). CCT consists of cognitively motivated, testable statistical models for various item formats, where the models posit parameters for the 'competence' of each informant as well as the 'culturally correct' answers to the questions. In addition, CCT models usually have parameters that reflect differential response biases of the informants and differential difficulty of the items. CCT models estimate competence, biases, and item difficulty endogenously along with the consensus answers and no prior (exogenous) calibration of the informants is imported into the evaluation. No theory of the structure of the shared information is assumed, and each informant is assumed to have an independent and unsystematic amount of the consensus knowledge. In addition, no axiomatic, information rules are imposed on the aggregation process because the desired 'pooled knowledge' is already specified by parameters in the model.

So far CCT models, along with associated inferential machinery, have been developed for the following item formats: dichotomous (true-false) items [4], [9], multiple choice items [20]; ranking [16]; matching [5]; presence or absence of ties in a digraph [2], Likert scale items [17], and items requiring a response in an interval [5]. Some models in CCT involve adapting a model in psychometric test theory that includes item scoring rules, and augmenting the model to make the scoring rules parameters in the model, for example one of the early papers in CCT is titled "Test theory without an answer key" [4]. In addition to pooling information from a single group of informants, CCT models are also applied in cross-cultural comparisons of beliefs [6], [17], [24].

The overall aims of a CCT model can be described as follows. Let $X_{ik}$ be a random variable that represents the response of informant $i$ to question $k$, and $\mathbf{X} = (X_{ik})_{NxM}$ the random *response profile matrix* for $N$ informants each answering the same set of $M$ questions. Ideally we want to use the information in a realization of $\mathbf{X}$ to:

a. Identify one or more latent "cultural groups" of informant that share information.
b. Decide if the statistical model used to do this provides an acceptable and valid account of the data; and, if so, for each cultural group to:
c. Estimate the "consensus answers."
d. Estimate the 'cultural competence' and 'response biases' of each informant
e. Estimate the difficulty of each item.

## 2 A CCT Model for Aggregating Digraphs

In order to illustrate CCT we adapt a model in [2] for the case of a decision maker (DM) who seeks to aggregate the judgments of multiple expert sources about the ties in a simple social network. A simple social network is represented by a digraph $S = <V, R>$ on a fixed set of $M>2$ nodes representing social entities, such as individual people, clubs, corporations, countries, and the like. The network is assumed to have a single social tie such as 'socializes with,' 'takes orders from,' 'conspires with,' 'shares information with,' or 'trades with.' The expert sources could be informants, data analysts, or even automated text processing algorithms. It is assumed that each of the $N$ sources provides a view of the social network, and since these views may differ the DM seeks a way to combine the experts' judgments into a consensus (most likely) social network along with a degree of belief in the validity of the consensus network.

### 2.1 Notation

Suppose each of $N$ sources reports dichotomous, presence/absence information about every ordered pair of distinct nodes in a digraph of $M$ nodes. Let the random variable $X_{i,jk} = 1$ if source $i$ indicates a tie from $j$ to $k$, otherwise $X_{i,jk} = 0$, and let the entire report of source $i$ be represented by the random matrix $\mathbf{X}_i = (X_{i,jk})$. The data from all sources consist of the three-way structure $\mathbf{X} = << X_{i,jk}>_{MxM}>_{i=1}^{N}$, and the DM seeks to aggregate the information in $\mathbf{X}$ to determine a 'consensus' (most likely) digraph $\mathbf{Z} = (Z_{jk})_{M \times M}$ , where the random variable $Z_{jk} = 1$ if there is a tie from $j$ to $k$, otherwise $Z_{jk} = 0$. In addition, the DM would like to assess the degree of evidence not only for the consensus digraph, but also for other less likely digraphs.

## 2.2 The Consensus Model

The basic model for aggregation regards the presence of a tie in the consensus digraph as a 'signal' and in addition to $\mathbf{Z}$ specifies two other classes of parameters, namely the hit and false alarm probabilities for each source at each ordered pair of distinct nodes, These are defined, as follows:

$$\forall 1 \leq i \leq N, \forall 1 \leq j, k \leq M, j \neq k, 0 < F_{i,jk} \leq H_{i,jk} < 1,$$

$$H_{i,jk} = \Pr(X_{i,jk} = 1 | Z_{jk} = 1), \ F_{i,jk} = \Pr(X_{i,jk} = 1 | Z_{jk} = 0).$$

The Model is specified by two axioms:

**Axiom 1.** (Common Truth). There is a fixed consensus digraph $\mathbf{Z}^* = (z_{jk}^*)_{M \times M}$ .

**Axiom 2.** (Conditional Independence). The reports from the sources satisfy conditional independence given for all realizations of $\mathbf{X} =<< x_{i,jk} >>$ by

$$\Pr[\mathbf{X} =<< x_{i,jk} >> | \mathbf{Z}^* = (z_{jk}^*)] = \prod_{i=1}^{N} \prod_{j=1}^{M} \prod_{\substack{k=1 \\ k \neq j}}^{M} \Pr(X_{i,jk} = x_{i,jk} | Z_{jk}^* = z_{jk}^*),$$

where $\Pr(X_{i,jk} = 1 | Z_{jk}^* = z_{jk}^*) = \begin{cases} H_{i,jk} & \text{if } z_{jk}^* = 1 \\ F_{i,jk} & \text{if } z_{jk}^* = 0 \end{cases}$.

Clearly the model in Axioms 1 and 2 is over parameterized as there are $M(M-1)$ parameters in $\mathbf{Z}^*$, $2NM(M-1)$ hit and false alarm parameters for the sources, and only $NM(M-1)$ bits in $\mathbf{X}$. Fortunately there are several ways to specify the model further to achieve parameter identifiability as follows for all $1 \leq i \leq N$ :

**Axiom 3a.** (Out-degree Homogeneity). $\forall 1 \leq j, k \leq M$, $H_{i,jk} = H_{i,j\bullet}, F_{i,jk} = F_{i,j\bullet}$ .

**Axiom 3b.** (In-degree Homogeneity). $\forall 1 \leq j, k \leq M$, $H_{i,jk} = H_{i,\bullet k}, F_{i,jk} = F_{i,\bullet k}$ .

Axiom 3a assumes that each source has identical hit and false alarm probabilities for all ties directed from any fixed node. However, these probabilities are allowed to vary from node to node within a source because each source may have differential amounts of knowledge in different regions in the digraph. Similarly, Axiom 3b assumes source homogeneity for all ties directed into any fixed node. It is of course possible to postulate both Axioms 3a and 3b; however, this would require that hits and false alarm probabilities, namely ($H_{i,..}, F_{i,..}$), be constant for each source over all nodes of the digraph, and this would lose the possibility that sources have differential knowledge.

It is easy to see that either of Axiom 3a or 3b, coupled with Axioms 1 and 2, allow the DM to consider information aggregation on a node by node basis. Consider Axiom 3a and pick a particular node $j$. Then the relevant information to aggregate in determining the ties from $j$ can be arrayed in the random matrix $\mathbf{R}_j = (X_{i,jk})_{N \times M}$.

Such a matrix has $M$-1 digraph parameters (since $z_{jj} \equiv 0$), $\mathbf{Z}_j^* =< z_{jk}^* >$, $2N$ hit and false alarm parameters, $\mathbf{H}_j =< H_{i,j\bullet} >, \mathbf{F}_j =< F_{i,j\bullet} >$, and $N(M-1)$ bits in $\mathbf{R}_j$. Thus as long as $N \geq 3, M \geq 4$, there are more bits in $\mathbf{R}_j$ than there are unknown parameters, so estimation is possible.

## 2.3 Parameter Estimation

We focus on estimating the parameters relevant to $R_j$, and given these estimates $\forall 1 \leq j \leq M$, estimation of $\mathbf{Z}^*$ can follow. One aspect of the model that makes parameter estimation not routine is that the answer key parameters are discrete. Several observations are useful.

**Observation 1.** Consider the model in Axioms 1, 2, 3a for a particular node $j$. Then the likelihood function is given by

$$L(\mathbf{H}_j, \mathbf{F}_j, \mathbf{Z}_j | \mathbf{R}_j) =$$

$$\prod_{i=1}^{N} \prod_{\substack{k=1 \\ k \neq j}}^{M} \left[ \frac{H_{i,j\bullet}(1 - F_{i,j\bullet})}{F_{i,j\bullet}(1 - H_{i,j\bullet})} \right]^{x_{i,jk} z_{jk}} \left[ \frac{1 - H_{i,j\bullet}}{1 - F_{i,j\bullet}} \right]^{z_{jk}} \left[ \frac{F_{i,j\bullet}}{1 - F_{i,j\bullet}} \right]^{x_{jk}} \left[ 1 - F_{i,j\bullet} \right].$$

**Observation 2.** Given observations of $\mathbf{H}_j, \mathbf{F}_j$, the values of $\mathbf{Z}_j^*$ that maximize the likelihood function are given by $\forall k = 1, ..., M, k \neq j, \hat{z}_{jk}^* = 1$ iff

$$\sum_{i=1}^{N} x_{i,jk} \ln \left[ \frac{H_{i,j\bullet}(1 - F_{i,j\bullet})}{F_{i,j\bullet}(1 - H_{i,j\bullet})} \right] \geq N \ln \left[ \frac{1 - F_{i,j\bullet}}{1 - H_{i,j\bullet}} \right].$$

**Observation 3.** Given $\mathbf{Z}_j^* = (z_{jk}^*)$, the values of $\mathbf{H}_j =< H_{i,j\bullet} >, \mathbf{F}_j =< F_{i,j\bullet} >$ that maximize the likelihood function are given by

$$\forall i = 1,\dots,N, H_{i,j\bullet} = \frac{\displaystyle\sum_{k=1}^{M} x_{i,jk} z_{jk}^*}{\displaystyle\sum_{k=1}^{M} z_{jk}^*}, \quad F_{i,j\bullet} = \frac{\displaystyle\sum_{k=1}^{M} x_{i,jk}(1 - z_{jk}^*)}{M - 1 - \displaystyle\sum_{k=1}^{M} z_{jk}^*},$$

provided $0 < \displaystyle\sum z_{jk}^* < M - 1$.

Observation 2 is particularly interesting because it shows that the CCT model aggregates responses from the expert sources by a log-odds weighting scheme. Such a scheme is an improvement over routine majority rule aggregation. Further, since the hit and false alarm parameters are determined endogenously, no external calibration of the expert sources is required.

There are several ways that these three observations can be useful in estimation. First from a classical perspective, one could seek M.L.E.s of the parameters by searching over the $2^{M-1}$ possible values of $\mathbf{Z}^*$ and using Observation 2. If $M$ is too large to allow this approach, then simulated annealing [1] can be applied as in [2] for a special case of this model. Alternatively, Bayesian estimation could proceed by specifying independent Bernoulli priors ($0<p<1$) on the elements of $\mathbf{Z}^*$ and independent beta priors on the elements of $\mathbf{H}_j$ and $\mathbf{F}_j$. Markov Chain Monte Carlo methods for this approach were used in [9] for a model for dichotomous responses similar to this one.

## 2.4 Interpreting Parameters

It is well known from the theory of signal detection [11] that hit and false alarm rates are difficult to interpret because they confound sensitivity to the signal and response bias. Once the hits and false alarm probabilities are estimated, one can reparameterize them by selecting from among a number of substantive signal detection models to interpret the competence and bias of the various sources in different regions of the digraph. For example, an application of the so-called single high threshold signal detection model would postulate a detection (competence) parameter, $D_{i,j} \in (0,1)$, for detecting the presence of a tie, and a response bias parameter, $g_{i,j} \in (0,1)$, for ascribing a tie when one is not detected. Then these new interpretable parameters are expressed in terms of the hit and false alarm probabilities, respectively, by

$$D_{i,j} = \frac{H_{i,j\bullet} - F_{i,j\bullet}}{1 - F_{i,j\bullet}}, \ g_{i,j} = F_{i,j\bullet}.$$

Also it is easy to accommodate other models of signal detection such as the traditional theory of signal detection (TSD) based on Gaussian distributions [2]. This model assumes that the presence of a tie (a signal) generates a strength of evidence on a one-dimensional decision axis that has a Gaussian distribution, and the absence of a tie (noise) also generates a Gaussian distributed strength but with a smaller mean. The difference of the means is specified by a parameter $d'_{i,j} > 0$, where larger values represent more competence by the source determining ties. The bias parameter $\beta_{i,j} > 0$ represents a threshold on the decision axis, where a tie is ascribed in case the evidence exceeds the threshold. Assuming the two Gaussians have equal variance, the new parameters are expressed in terms of hits and false alarms by

$$d'_{i,j} = \Phi^{-1}(H_{i,j\bullet}) - \Phi^{-1}(F_{i,j\bullet}), \ \beta_{i,j} = -(1/2)[\Phi^{-1}(H_{i,j\bullet}) + \Phi^{-1}(F_{i,j\bullet})].$$

## 2.5 Possible Extensions

Possible extensions of the CCT model for digraph aggregation are to estimate the consensus digraph subject to constraints. The constraints could take the form of discrete axioms (e.g. linear orders, semi-orders, taxonomies, and predetermined social network structures) or probabilistic assumptions. If discrete axioms are imposed apriori on the consensus digraph, one wants to aggregate data from expert sources to consider the most likely digraph subject to the imposed axioms. On the surface, such a project would seem to be infeasible because there is not a clear cut way to search the combinatorially explosive space of all binary relations on a set of nodes subject to axioms while skipping over the majority of binary relations that do not satisfy the axioms. A key insight to this extension comes from the work of J.C. Falmagne and colleagues, e.g. [7] on well graded relational structures. In a nutshell, the idea is that certain axiomatic relational systems have a property called *well gradedness* which implies that one can move from model to model of the axioms by deletion or addition of a single ordered pair (e.g. this is obvious for linear orders). In such cases one could select a quantity reflecting the 'fit' (or likelihood) of a model and move using simulated annealing [1] by small steps through all models of the axioms to find a digraph satisfying the axioms that optimizes the measure.

This proposed approach of imposing discrete axioms on consensus relational structures provides a probabilistic alternative for a standard area in applied discrete mathematics sometimes called algebraic consensus. In this area, several sources each provide a model of some digraph structure on a set of nodes such as a lattice, tree, or median graph, and an algebraic consensus structure is determined. For example, if a set of informants each suggest a binary tree on a set of $M$ points, an algebraic consensus structure might be a tree at the 'median' of their trees, e.g. [12], [13]. The algebraic consensus approach in this area is distinctly non probabilistic, and the CCT approach may overcome the fact that algebraic consensus structures are quite fragile to small changes in the informants' structures that are aggregated.

One way to impose probabilistic constraints on the consensus digraph is to modify Axiom 2. All models of CCT employ a conditional independence axiom that states that the individual source-question responses are conditionally independent given parameters that describe the consensus answers. In cases where the questions are unrelated to each other, the axiom is formulated like that of Axiom 2, where the only information needed to specify conditional response independence is the consensus answer to the relevant question. However, when questions are arguably interrelated, then more information about the consensus may be needed to achieve conditional independence. One example is the CCT model for a matching questionnaire [5], where matching items to question stems is without replacement.

In the case of social networks, it is quite reasonable to expect that a source's response to a particular tie will depend on the relevant structure of the consensus network, especially its local structure. It is straightforward to reformulate Axiom 2 as follows:

$$\Pr[\mathbf{X} = << x_{i,jk} >> | \mathbf{Z}^* = (z_{jk})] = \prod_{i=1}^{N} \prod_{j=1}^{M} \prod_{\substack{k=1 \\ k \neq j}}^{M} \Pr(X_{i,jk} = x_{i,jk} | A_{jk}^*),$$

where $A_{jk}^*$ is a subset the ties in $\mathbf{Z}^*$ that may influence the response probability of $X_{i,jk}$. One natural possibility is to make a Markovian assumption as in [19], [20]. In this case the subset of ties that can influence the response probability of $X_{i,jk}$ are ties that share at least one node with $\{i, j\}$, namely $A_{jk}^* = \{z_{uv}^* | \{u, v\} \cap \{i, j\} \neq \varnothing$. Then conditional response probabilities for each possible pattern of ties in $A_{jk}^*$ would need to be specified in terms of parameters reflecting each source's competence and bias.

The Markovian extension involves adding probabilistic constraints that depend on several relevant ties in the consensus digraph. Another class of probabilistic constraints would be to condition on latent variables that specify the location of the nodes in a metric space. In the case of covert social networks, this may be an important addition because there may be an underlying metric of social closeness, and closeness may increase the probability of a social tie, e.g. [10]. There are several probabilistic models of social networks that have this structure, e.g. [8].

# References

1. Aarts E, Korst, J (1989) Simulated annealing and Blotzman Machines. Wiley, New York

2. Batchelder W H, Kumbasar E, Boyd J P (1997) Consensus analysis of three way social network data. J Math Sociol 22:29-58
3. Batchelder, W H, Romney, A K (1986) The statistical analysis of a general Condorcet model for dichotomous choice situations. In: Grofman B, Owen G (eds) Information pooling and group decision making: Proceedings of the Second University of California Irvine Conference on Political Economy. JAI Press, Greenwich, Conn
4. Batchelder W H, Romney A K (1988) Test theory without an answer key. Psychometrika 53:71-92
5. Batchelder W H, Romney A K (1989) New results in test theory without an answer key. In: Roskam E E (ed) Mathematical Psychology in Progress. Springer-Verlag, Heidelberg
6. Batchelder W H, Romney A K (2000) Extending cultural consensus theory to comparisons among cultures. Mathematical Behavioral Sciences Technical Report No. 00-17, Institute for Mathematical Behavioral Sciences, University of California Irvine
7. Doignon J P, Falmagne J C (1997) Well graded families of relations. Discrete Math 173:35-448.
8. Hoff P D, Raftery A E, Hancock M S (2002) Latent space approaches to social network analysis. J Am Stat Assoc 97:1090-1098
9. Karabatsos G, Batchelder W H (2003) Markov chain estimation theory methods for test theory without an answer key . Psychometrika 68:373-389
10. Lazarsfeld P F, Merton R K (1964) Friendship as social process: A substantive and methodological analysis. In Berger M, Abel T, Page C H (eds) Freedom and control in modern society. D. Van Nostrand, New York
11. McMillan N A, Creelman C D (2004) Detection theory a users guide, 2nd edn. Cambridge University Press, New York
12. McMorris F R, & Neumann D (1985) Axioms for consensus functions on undirected phylogenetic trees. Math Biosci 74: 17-21
13. McMorris F M, Mulder H M, Roberts F M (1998) The median procedure on median graphs. Discrete Appl Math 84: 165-181
14. Romney A K (1999). Cultural consensus as a statistical model. Curr Anthropol 40, Supplement: S103-S115
15. Romney A K, Batchelder W H (1999) Cultural consensus theory. In Wilson R, Keil F (eds) The MIT Encyclopedia of the Cognitive Sciences. The MIT Press, Cambridge, MA
16. Romney A K, Batchelder W H, Weller S C (1987) Recent applications of cultural consensus theory. Am Behav Sci 31: 129-149
17. Romney A K, Moore C C , Batchelder W H, Hsia T (2000) Statistical methods for characterizing similarities and differences between semantic structures. P Natl Acad Sci 97: 518-52
18. Romney A K, Weller S A, Batchelder W H (1986) Culture as consensus: A theory of culture and informant accuracy. Am Anthropol 88: 313-338
19. Strauss D, Ikeda M (1990). Pseudolikelihood estimation for social networks. J Am Stat Assoc 85: 204-212
20. Wasserman S, Pattison P (1996) Logit models and logistic regression for social networks I. An introduction to Markov random graphs and p*. Psychometrika, 60: 401-426
21. Weller S A (2007) Cultural consensus theory: Applications and frequently asked questions. Field Methods 19: 339-368

# Dynamic Networks: Rapid Assessment of Changing Scenarios

Nadya Belov[†], Michael K. Martin[*], Jeff Patti[†], Jeff Reminga[*], Angela Pawlowski[†] and Kathleen M. Carley[*]

[†]<nbelov, jpatti, apawlows>@atl.lmco.com, Lockheed Martin Advanced Technology Laboratories, Cherry Hill, New Jersey
[*]<mkmartin, reminga, kathleen.carley>@cs.cmu.edu, Center for Computational Analysis of Social and Organizational Systems, Institute for Software Research, Carnegie Melon University, Pittsburgh, Pennsylvania

**Abstract** As events unfold, the underlying networks change. Most network science tools, however, assume analysts have a single snapshot of the data, or at most, a second snapshot at a different time. The underlying network representation schemes, assessment technologies, and visualizations do not lend themselves naturally to dynamic networks. Herein, we identify key criteria for network representation of dynamic, uncertain information and present a technology enabler that combines information fusion and dynamic network analysis. The value of technological solutions for network analytics in a dynamic environment are discussed.

## 1 Introduction

New challenges have arisen as network scientists move to apply their analytical techniques to military and other real-world environments. The target environments involve large amounts of relational data that characterize dynamic, often distributed, decision-making situations. These situations require time-sensitive network analyses to support command decision-making as real world events alter the networks and, sometimes, prudent courses of action. Analysts need to capture and assess the impact of these changes, with minimal computational overhead; they require Network Representation Languages (NRLs) that adequately support the continuous ingestion of battlefield data, kinetic and non-kinetic, into dynamic network analysis packages such as ORA [1]. In the following, we identify core requirements for NRLs to meet this need, and sketch the value of rapid network assessment.

We believe the core requirement for continuous network analysis of battlefield data is an NRL that adequately handles dynamic, uncertain meta-network data. For many xml schemes — such as GraphML — it is possible to do all the things we are suggesting in terms of edges and meta-data. However, that does not mean the approach is easy or there is a network analysis tool that can understand that information. Our goal is to

H. Liu et al. (eds.), *Social Computing and Behavioral Modeling*,
DOI: 10.1007/978-1-4419-0056-2_6, © Springer Science + Business Media, LLC 2009

identify what changes are needed to handle real time data, extending DynetML to produce CoreNetML and DyNetML 2.0. We further extend our goal to re-use tools such as ORA, originally design for static analysis, to assist in dynamic network analysis. In this paper we discuss how this is accomplished through CHANS. The majority of well-known NRLs (e.g., CSV, edge-list, UCINET-native [2], and GraphML [3] are very good at representing static networks. The representation of uncertain observations, however, can be difficult if not impossible. Three key improvements are needed to handle the battlespace: Enhanced Edge Representation (EER), Enhanced Node Representation (ENR), and Enhanced Meta-Data Representation (EMR).

In Section 2, we define criteria for network representation of distributed, dynamic, uncertain environments, and provide a high level score card that compares representation capabilities of well-known NRLs, alongside enhanced representations such as CoreNetML—developed by Lockheed Martin Advanced Technology Laboratories (LM ATL) as an extension of DyNetML 1.0 (see [4] for an initial description of DyNetML and http://www.casos.cs.cmu.edu/projects/dynetml/ for the 1.0 specification). Next, in Sections 3 and 4, we illustrate how enhanced NRLs can enable dynamic battlefield assessment in a notional system where battlefield observations are input to LM ATL's Core HumAn Network System (CHANS) prototype and updated DyNetML is exported to ORA for rapid analysis of network changes. Section 5 concludes.

## 2 Criteria for Dynamic Distributed Network Analytics

The dynamic, distributed, uncertain character of real world settings often yields incomplete data. For example, examination of the Tanzania embassy bombing indicates the data arrived piecemeal. A group of terrorists worked together to build, deploy and explode a bomb. Early attention to this group might have revealed some work related to bomb manufacturing, but only as the events unfolded would information start appearing on associated surveillance and transportation activities. Even as data became available, it may not have been reliable or complete because it would have depended on leads from informants, tracking via sensors, and so on.

Data from real situations has these features: 1) one or more sources with characteristics such reliability and pedigree, 2) potential delays in positive identification of unknown nodes (entities), and 3) nodes and edges (relations) that may appear, disappear, or change in strength over time as new information arrives. In addition, the networks of concern are "meta-networks" that are multi-modal (different types of nodes, e.g., people, places, assets), multi-link (different types of relations), and multi-level (multiple types of networks, e.g., variations in the granularity of representation such as individuals versus groups).

To analyze such data, a simple solution is to divide it into time periods and then assess a sequence of networks, each defined at a particular level of reliability. However, from a computational perspective such a solution does not scale. Computationally, the key criteria are: 1) scalability, 2) minimization of data IO costs, 3) facilitation of temporal analysis, and 4) facilitation of sensitivity analysis.

Real-world data characteristics and computational criteria imply a need for attributed data. Three types of attribute information need to be represented:

1)  Edge Attributes: Details including strength, direction, reliability, existence, and change in these factors (DyNetML 1.0 with EER).

2)  Node Attributes: Details including existence, attributes about the state of the node such as level of education or age, and so on (DyNetML 1.0 with ENR).

3)  Meta-Data Attributes: Details about who generated the data, why, when, how, coding constraints, information pedigree, credibility, level of classification, etc. (DyNetML 1.0 with EMR).

CoreNetML, the LM ATL enhancement of DynetML 1.0, addresses the features of real-world data outlined above by encoding varying and often conflicting information, along with temporal and geo-spatial markers; encoded confidence values based on source and pedigree identifiers enable de-confliction by outside data normalization services. CoreNetML exemplifies aspects of EER, ENR, and EMR in the representation of real-world data. In a similar vein, a new DyNetML 2.0 with EER, ENR, and EMR as standard features, and a new version of ORA using these features, will be released November 2008.

Table 1 scores common NRLs, alongside enhanced NRLs, in terms of their capability to represent attributed data. For each feature the NRL is rated as √=yes, D=difficult, or U=unknown/to be determined; a blank cell means no. Additionally, √*, means that when there are multiple edges between the same pair of nodes, an equivalent way of handling classes of such edges is with attributes on the network for that edge class rather than on the edge itself. This form of representation is more compact.

As can be seen in Table 1, the various enhancements each focus on a section of the feature-set of a network. A wide range of capabilities are now enabled by attribute-enhanced representations (e.g., CoreNetML, DyNetML 2.0). In the area of network dynamics, these include, but are not limited to:

1)  Ingestion and assessment of real time changes.

2)  Characterization of change in groups by change in members attributes.

3)  Visualization of and detection of change in networks with minimal IO.

4)  Tighter link to simulation and forecasting tools as ability to handle evolving networks including node attributes such as age and beliefs are possible.

5)  Ability to reason about, locate patterns in, and track trail data over time.

Table 1. Score-Card of existing network representations highlights the necessary features provided by CoreNetML and DyNetML 2.0

| Representational Capability | Edge Lists | CSV | UCINET – native | GraphML | DyNetML 1.0 | CoreNetML | DyNetML with EER | DyNetML with ENR | DyNetML with EMR |
|---|---|---|---|---|---|---|---|---|---|
| Scalable to Large Datasets | | | | √ | √ | U | U | √ | √ |
| Nodes – Multi-mode | | D | | √ | √ | √ | √ | √ | √ |
| Meta-ontology | | | | | √ | √ | √ | √ | √ |
| Node Attributes | | | | | | | | | |
| Name | √ | √ | | √ | √ | √ | √ | √ | √ |
| ID | √ | √ | | √ | √ | √ | √ | √ | √ |
| String | | | | √ | | √ | | √ | |
| Dates | | | | | | √ | √ | √ | |
| Data type | | | | | | | | √ | |
| Numerical | | | | | √ | √ | √ | √ | √ |
| Edges – Multi-link | | D | √ | √ | √ | √ | √ | √ | √ |
| Edge Attributes | | | | | | | | | |
| String | | | | √ | √* | √ | √ | √* | √* |
| Numerical | | √ | √ | √ | √ | √ | √ | √ | √ |
| Temporal tagging | | | | | | √ | √ | | |
| Temporal deltas | | | | | | | | | |
| Meta-network Meta-data | | | | | | | | | |
| Coding choices | | | | | | | | | √ |
| Source | | | | √ | √ | | | √ | √ |
| Source attributes | | | | | | | | | √ |
| Date | | | | √ | √ | √ | √ | √ | √ |
| Author | | | | √ | | | | | √ |
| Other context | | | | | √ | √ | √ | √ | √ |
| Multi-level | | D | | √ | √ | √ | √ | √ | √ |
| Network attributes | √ | √ | √ | √* | √* | √* | √* | √* | √ |
| Name | √ | √ | √ | √ | √ | √ | √ | √ | √ |
| ID | √ | √ | √ | √ | √ | √ | √ | √ | √ |

## 3 Enhanced NRLs and Dynamic Battlefield Analysis

Battlefields require timely collection and assessment of new information in the context of prior information. This new information may be confirmations of suppositions, completely new intelligence, reports on the impact of a course of action and so on. From a network standpoint, intelligence changes the network by altering the attributes of nodes or edges and the presence of nodes or edges. Attribute-enhanced NRLs support the tracking of such changes, and open possibilities for dynamic battlefield assessment systems comprised of dynamic network analysis technologies. Information fusion of soft, non-kinetic and potentially conflicting data has only recently been recognized as a core requirement for dynamic network analysis [5]. Figure 1 illustrates such a system using Core HumAn Network System from LM ATL, and ORA and Construct from CMU.

**Figure 1.** Attribute-enhanced NRLs facilitate network analysis of newly observed changes and prediction of expected changes using simulation tools such as Construct [6].

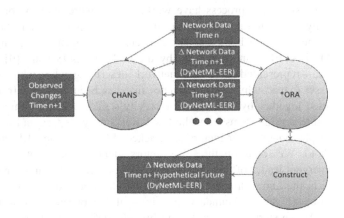

CHANS is designed to bring together disparate information, data sources and network analysis technologies to produce effective classification and forecasting results. It includes a central data store called the Human Network Core (HNC), where all human network and communication data is persisted, a Data Fusion Service responsible for maintaining data integrity, and Forecasting and Classification Interfaces that provide network analysis tools with access into the HNC.

The HNC is central in the CHANS. It persists data represented in CoreNetML from various sources including, but not limited to imagery, human generated reports and open data sources such as blogs and news feeds. The Data Fusion Service addresses significant hurdles to effective network analysis of evolving battlefield situations — data normalization — by assessing confidence in the accuracy of incoming data based on its source, pedigree, occurrence and receipt time, and then normalizing incoming data to reflect confidence based on existing data.

CHANS provides interfaces that facilitate the retrieval of information stored in the HNC for use by Forecasting and Classification services such as ORA and Construct. Edge enhancement in CoreNetML, for example, facilitates faster ingest, and assessment due to reduced IO, for both observed and forecasted changes. Consequently, techniques for dynamical analysis such as network change detection—a statistical process control approach to detecting substantive changes in networks—can be applied to real and simulated data ([7] [8], respectively). Expected changes due to network adaptation and healing can be forecast using simulation tools such as Construct [6].

# 4 Scenario

How would the process have worked if we were able to conduct dynamic network analytics at the time of the Tanzania bombing in 1998? To illustrate, we use a subset of the Tanzania bombing data collected from open sources by Connie Fournelle at ALPHATECH and expanded by the CMU CASOS team [9]. We can think of this data—represented as a meta-network describing who, what, where, how and, to a lesser extent, why—as intelligence that has been collected from diverse sources and fused together by CHANS. For brevity's sake, we show three periods of change in Figure 2 using items from the ORA Key Entity Report, and in the images show only people though other factors are tracked. El Hage is an FBI photo; Owhali is a CNN photo. We focus on three key actors identified in ORA's Key Entity Report: the emergent leader of the group (highest in cognitive demand), the person most likely to be in the know (highest in degree centrality), and the person most likely to connect groups (highest in boundary spanning—high betweenness, low degree centrality).

As we start the scenario, the brigade staff has been using CHANS and ORA to continuously update their understanding of the evolving situation. Late 1996: trusted informant warns al Qaeda may attack embassies; known cell structure is in Figure 2 as Time 1. Analyses show Khalfan Mohamed to be in-the-know and the emergent leader; Abdullah Ahmed Abdullah is the boundary spanner. February 1997: Wadih el Hage returns to Kenya and phone monitoring picks up el Hage in discussions with terror cell. This is picked up and processed by CHANS. Network analyses show a shift in key actors, with el Hage identified as in-the-know and the emergent leader, and Mohammed Odeh as the boundary spanner. May 1998: image analysis reveals that Mohamed Owhali is connected to Bin Laden—leading to another shift in structure as the group nears the August bombing, with el Hage in-the-know, Khalfan Mohamed again as the emergent leader, and Owhali as the boundary spanner.

At each time period, CHANS ingests new reports, fuses these reports with the existing network using the Data Fusion Service and produces a revised DyNetML that

**Figure 2.**
Change in
Networks.

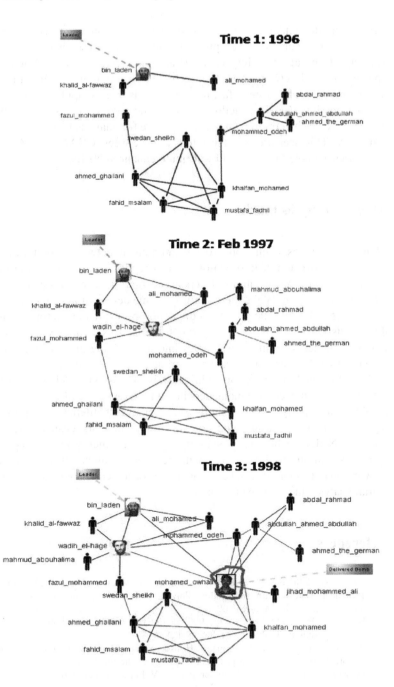

reflects reported changes in the network. This is imported into ORA, and the results are immediately compared with previous data. As new observations arrive, the relative standing of key actors is impacted — as are the implied courses of action. Until now updates to social networks based on incoming information were in a time consuming, manual manner. Therefore, a fast impact analysis due to rapid changes in the network was not possible. CHANS facilitates timely and automated fusion of newly available data into a cohesive, up-to-date network picture for analysis by tools such as Construct and ORA. This scenario and the interaction between CHANS, ORA and Construct showcases the potential enabled by attribute-enhanced NLRs.

## 5 Concluding Remarks

CHANS facilitates continuous social and human network monitoring through the collection and fusion of information such as reports from observers in the battlefield who report on events, locations of persons of interest, suspicious containers, and other relevant information. While there is always some degree of uncertainty associated with the reports, CHANS attempts to address this by estimating information confidence through its Data Fusion Service. Through the Core Network System component, the CHANS merges incoming observables into an evolving network representation. The outputs of CHANS are an updated network that can be readily imported into a network analysis tool such as ORA for rapid analysis of substantive changes in the network, which can be fed back to commanders in the field on short order. CHANS provides the capability to rapidly ingest, assimilate, and analyze battlefield and provide feedback regarding important current and potential future network changes. Modifications to standard NRLs enable this capability by enhancements to the representation of dynamics and uncertainty, along with reduced computational overhead. Dynamic network analytics can then take advantage of observed changes in the data to engage in trail analysis and change detection.

## References

1. Carley K, Columbus D, DeReno M, Reminga J, Moon I (2008) ORA User's Guide 2008. Technical Report, Carnegie Mellon University, School of Computer Science, Institute for Software Research, CMU-ISR-08-125.
2. Borgatti S, Everett M, Freeman L (1999) UCINET 5 for Windows: Software for Social Network Analysis. Analytic Technologies, Inc., Natick, MA.
3. Brandes U, Eiglsperger M, Herman I, Himsolt M, and Marshall M.S (2002) GraphML Progress Report: Structural Layer Proposal, Proceedings of the 9th International Symposium on Graph Drawing, Lecture Notes of Computer Science, Volume 2265, pp. 501-512.

4.  Tsvetovat M, Reminga J, Carley K (2004) DyNetML: Interchange Format for Rich Social Network Data. CASOS Technical Report. Carnegie Mellon University, School of Computer Science, Institute for Software Research International, CMU-ISRI-04-105.
5.  Nick Morizio (2008) Context Selection for Linguistic Data Fusion, in the Proceedings of the 11th International Conference on Information Fusion, Cologne, Germany, June 30-July 3, 2008.
6.  Schreiber C, Carley K (2004) Construct - A Multi-agent Network Model for the Co-evolution of Agents and Socio-cultural Environments. Construct - A Multi-agent Network Model for the Co-evolution of Agents and Socio-cultural Environments. Technical Report, Carnegie Mellon University, School of Computer Science, Institute for Software Research International, CMU-ISRI-04-109.
7.  McCulloh I, Carley K (2008a) Social Network Change Detection. Technical Report, Carnegie Mellon University, School of Computer Science, Institute for Software Research, CMU-ISR-08-116.
8.  McCulloh I, Carley K (2008b) Detecting Change in Human Social Behavior Simulation. Technical Report, Carnegie Mellon University, School of Computer Science, Institute for Software Research, CMU-ISR-08-135.
9.  Moon I, Carley K (2007) Modeling and Simulation of Terrorist Networks in Social and Geospatial Dimensions. IEEE Intelligent Systems, Special issue on Special issue on Social Computing - Sep/Oct '07, 22, 40-49.

# Acknowledgements

The work by CMU on DyNetML 2.0, *ORA and Construct, was supported in part by the Air Force Research Laboratory (AFRL) CAPES Project: FA865007C6769, the Department of Defense Multi-disciplinary University Research Initiative (MURI) in cooperation with George Mason University (GMU) and Air Force Office of Scientific Research (AFOSR): 600322 and the Office of Naval Research (ONR) DNA: N00014-06-1-0104. Additional support for *ORA was provided by the center for Computational Analysis of Social and Organizational Systems (CASOS) and the Institute for Software Research at Carnegie Mellon University. The views and conclusions contained in this document are those of the authors and should not be interpreted as representing the official policies, either expressed or implied, of the United States Air Force, United States Department of the Navy, U.S. Department of Defense, or the U.S. government.

# Modeling Populations of Interest in Order to Simulate Cultural Response to Influence Activities

Michael Bernard[a], George Backus[b], Matthew Glickman[c], Charles Gieseler[d], and Russel Waymire[e]

{mlberna[a], gabacku[b], mrglick[c], cjgiese[d], and rlwaymi[e]}@sandia.gov, Sandia National Laboratories, Albuquerque, New Mexico

**Abstract** This paper describes an effort by Sandia National Laboratories to model and simulate populations of specific countries of interest as well as the population's primary influencers, such as government and military leaders. To accomplish this, high definition cognition models are being coupled with an aggregate model of a population to produce a prototype, dynamic cultural representation of a specific country of interest. The objective is to develop a systems-level, intrinsic security capability that will allow analysts to better assess the potential actions, counteractions, and influence of powerful individuals within a country of interest before, during, and after an US initiated event.

## 1 Societal Assessment Capability

The United States is finding itself increasingly engaged in the development of unconventional partnerships that require a variety of non-traditional activities to better support political and economic stability in regions of interest. Unfortunately, there is no effective means to adequately forecast and assess how both individual leaders, and the people they influence, will behave with regard to possible US policies and actions. It is asserted here that an accurate characterization of a society must represent this interaction between people under control, those influencing power, and external variables, such as US actions or oil revenue variation (in counties dependent on oil). While assessment tools have modeled and simulated societies, they have, thus far, been limited to gross behavioral models. Furthermore, no existing macroeconomic or societal model addresses security dynamics or coordinated multiple kinetic and non-kinetic interventions. We believe that the phenomena that maintain or transition dictatorship and democracy have recently become understandable enough to pose testable hypotheses amenable to simulation. As such, the ability to address intervention dynamics and unintended, higher order consequences is a key goal of this work. In pursuit of this goal, Sandia National Laboratories (Sandia) has developed a prototype societal assessment capability that assists in the behavioral influence analysis of foreign targets

H. Liu et al. (eds.), *Social Computing and Behavioral Modeling*,
DOI: 10.1007/978-1-4419-0056-2_7, © Springer Science + Business Media, LLC 2009

of interest. The objective of the described work is to develop a systems-level capability that will allow analysts to better assess potential actions and counter-actions of individuals interacting within a foreign country of interest before, during, and after an US initiated event. The assessment is designed to address the dynamics that drive stability and instability. Specifically, it is designed to: (1) assess adversarial choice options that allow analysts to pose "what-if" queries concerning hypothetical policy and/or military initiatives to help determine how and why a population may react to a specific event, leader, or operation across time, (2) assess potential blind spots by providing analysts with the ability to better understand higher order interaction effects between leaders and local societies and how allegiances are formed and changed over time, (3) perform risk analysis by determining the limiting assumptions and unknowns for the successful outcome, and (4) perform risk management by establishing whether there are delayed consequences that will require mitigation or adjustments to planning. Collectively, this type of simulation is designed to permit assessment of shaping activities and US tactics in an operational environment by creating a system that can help an analyst better understand the interaction between leaders and local societies and how allegiances are formed and changed over time.

To accomplish this Sandia is utilizing its extensive technical expertise in Modeling & Simulation (M&S) to create a social simulation platform that couples High-Definition Cognitive Models (HDCM) with a cultural, economic, and policy-based simulation. The HDCMs are purposely designed to computationally represent the mindset of specific individuals, including their cognitive perceptions, goals, emotion states, and action intentions. The actions of one HDCM can affect the mindset and actions of others, as well as the general mindset of the society in which they are situated. The society, computationally represented in this initial effort by Sandia's Systems Dynamics-based Aggregate Societal Model (SDASM) can, in turn, affect the actions of the HDCMs (see Figure 1). The HDCM is focused on individual or small-group level of analysis, whereas the SDASM is focused at an aggregate level social, economic, and cultural level of analysis. These models are joined to provide a high-fidelity, scaleable assessment tool of individuals, small groups, and society to produce outcome distributions investigating attitudinal and behavioral reactions to US policies for a given country, group, or ethnic region.

Figure 1. A conceptual view of Sandia's High Definition Aggregate Societal Modeling Framework

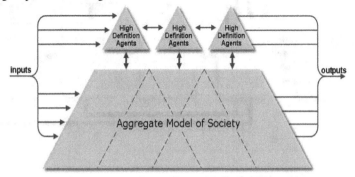

43

## 2 High-Definition Cognitive Models

As stated, the HDCM agents can represent the goals, cognitive perceptions, and dynamic emotion states of specific individuals or types of individuals that interact with each other and their environment in a psychologically plausible manner [3, 8]. To achieve this, the HDCMs are based on robust psychological research and theory. Accordingly, an important feature of the framework is its primary emphasis on psychological realism. In this way several different types of individuals (e.g., government or group leaders) can be represented who may be generally similar to one another, but who exhibit differences in attitudes and behavior. This realism is achieved by two means. First, the underlying cognitive processes that are modeled are based on recent advances in cognitive neuroscience, decision theory, and sociology. Second, HDCMs are data driven. Data can come from subject matter experts, intelligence reports, the media, as well as other sources. The HDCMs consist of a human-representative computational model through which it recognizes patterns of stimuli in the environment and responds to those stimuli according to current contexts. As an HDCM agent perceives its environment, its perceptions are influenced by a hierarchy of higher-level goals or moral states, as well as emotion states (see Figure 2). The goals and perceptions together trigger specific intermediate goals for activation. Depending on the current stimuli, specific intermediate goals will become activated. This combined with emotion states produce an action intention. The most activated intention becomes a possible action.

Figure 2: The process diagram of the cognitive actions within the HDCM framework

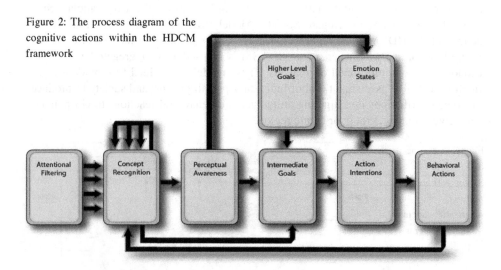

The behaviors associated with possible actions conform to the theory of planned behavior, which maintains that behaviors are influenced by attitudes towards a specific behavior, the subjective norms associated with acting out that behavior, and the perception that this behavior is within a person's control. This forms an action intention state, which then typically drives that person's actual behavior [1, 7]. This type of high-fidelity representation can capture and express the basic psychological processes of individuals (e.g., leaders, terrorists). A key component to this technology is its computational model framework whereby a modeled human recognizes patterns of stimuli in the environment and responds appropriately to those stimuli according to prior experiences via its semantic knowledge and pattern recognition modules. The semantic module incorporates an associative network with nodes representing the critical concepts or "schemas" in an agents "mind." The pattern recognition and comparator modules can provide mechanisms for: (1) evaluating the evidence provided by cues favoring or conflicting with each situation and (2) implementation of top-down activation. Implicit to recognition of a situation, there is recognition of goals, or attainable states, and the actions needed to realize those goals, including likely intermediate states. The cognitive subsystem in our model serves as the point where the diverse emotions, memories, and other factors of an individual are all used to generate a decision for action (or inaction). Actions of the individual and their repercussions then effect how the aggregate model transitions to the next state (or return to the same state). At present, the emotions this cognitive model represents are anxiety-fear and frustration-anger. Activation of a specific concept or situation produces activation of associated emotional components.

## 3 Systems Dynamics-based Aggregate Societal Model

The SDASM consists of a calibrated, systems dynamics/socio-political framework with behavioral decision simulation within populations and governments. It incorporates cultural, institutional, economic, and political distinctions. SDASM includes logic for detailed intra- and inter-regional interactions, as well as aggregate rest-of-world feedback dynamics. A calibrated framework combines selected economic data and societal index sources to allow model parameterization and long-term global modeling capability. Currently, no existing macroeconomic or societal model addresses security dynamics or coordinated kinetic and non-kinetic intervention. Methods developed at Sandia, combined with new verification and validation approaches under development at Sandia can, however, provide robust behavioral-response simulations [2, 12]. The foundation of these methods come from Nobel Prize winning work of Daniel McFadden on Qualitative Choice Theory (that accurately portrays human decision making) and by Clive Granger on Cointegration (that determines those variables which affect decisions with enduring or transient significance).

The physical and economic behavioral implications are readily simulated using basic aspects of conventional simulation methods such as System Dynamics [14], engineering [9], and economics [11]. Societal and economic realities are the consequence of behavioral decisions. The simulation and understanding of these processes is only recently possible. Decisions are the process of making choices. All behaviors are the consequence of choices made. McFadden pioneered the use of (psychologically framed) qualitative choice theory (QCT). QCT [13] is actually very quantitative and determines the importance people place on information, tastes, beliefs, and preferences when making decisions. The robust parameterization of QCT is often based on data readily obtainable in the field. Other techniques can further determine the correct functional representation of the QCT utility formulation for the problem at hand [10].

A key part of the decision process is the filtering of information and the extent to which experience biases the decision process. At a group level, the probabilistic nature leads to a mean-value response because random variation in one direction by one person is balanced by the reverse variation of another person. The enduring aspects of the population (society) dominate the group behaviors. The identification of the transient and stable components of the decision process use cointegration (also Granger Causality) methods pioneered by Granger. These same methods also ascertain the filtering and delayed-response processes associated with information perception and behavior [5, 6]. These methods and others are summarized in Backus at al. [2] and [4]. These techniques can integrate disparate perspectives and information, qualitative as well as quantitative, into analysis and decision support systems. The methods are compatible with orthodox macroeconomic assumptions and used for all matter of choices (including those associated with security).

## 4 Societal Assessment Prototype

Applying the techniques and models discussed above, Sandia has produced a prototype societal assessment capability that shows (1) potential actions, as well as the psychological processes behind those processes, for specific individuals of interest; and (2) potential societal actions in response to the actions of individuals of interest as well as exogenous variables. In the system, the inputs to the HDCM/SDASM system are cues associated with environmental events, US actions, and other external forces. These cues can be actual events, or be posed by analysts to create "what-if" scenarios. The cues will affect the HDCM by creating perceptions that are particular to a specific HDCM agent. The resulting cognitive states and actions will serve as inputs to the SDASM. The SDASM will represent the society in which the HDCM agents wield influence. The SDASM will receive the same cues as the HDCMs, as well as other cues that affect societies at an aggregate level. The output of the SDASM will serve as additional cues to the HDCMs.

When fully implemented, it is believed the combined interactions will capture the dynamics, secondary effects, and potential unintended consequences so as to better assess/develop interventions and regional-stabilization conditions. Figure 3 shows an example of this process for a single individual as well as the interaction between the individual and the societal model. Incoming information activates specific concepts (shown in red) to represent specific modeled psychological processes (such as perceptions and goals). Potential actions will be fed to the SDASM, which will, in turn, activate concepts that will be fed to the HDCMs. The interactions from this process are then visualized in a graphical interface.

Figure 3. An example of the output of the prototype societal assessment tool

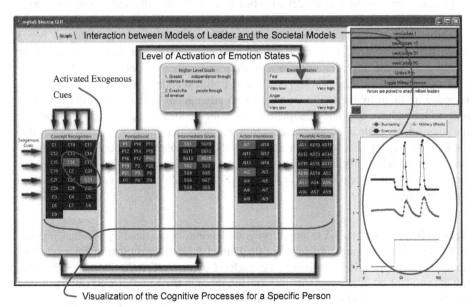

## 5 Conclusion

We believe this modeling exploration provides evidence for the potential value and viability of combining cognitive models to represent individual leadership with System Dynamics models to simulate groups and societal interactions. The demonstration model also shows it is possible to design a model that does allow field data for parameterization (and thereby allows validation testing/modification) of the model. While framework discussed here is intriguing, the use of normalized parameters and unsubstantiated assumptions means that there is, as yet, no legitimacy to quantitative results.[1] Confidence in model results/recommendations would require client supported data efforts, Subject Matter Expert review, and formal model validation and verification. The initial prototype uses a simpler societal model to explore and gain understanding of the problem domain. A US policy or other directives proposed by the sponsor will be modeled to forecast change in aggregate behavior. This will permit large-scale simulations that model societies reacting to hypothetical US actions prior to an engagement.

Ultimately, this capability could make significant headway in the ability to better understand and forecast attitudinal and behavioral responses at a regional, national, or local level. Specifically, the attitudes and actions of a given population would be modeled and simulated before, during, and after a political and/or military action has been imposed on them by the US or its allies. Past attempts at assessing conflict initiation and evolution have depended on quantifying static conditions, such as poverty or ethnic majorities—with minimal success. New methods that address the behavioral dynamics and expectation formation appear to show much promise. Enhancing macro-economic models to include endogenous security metrics and adding behavioral dynamics should produce a reliable tool set that Sandia and the nation can use to address emerging and evolving threats.

**Acknowledgements** Sandia is a multiprogram laboratory operated by Sandia Corporation, a Lockheed Martin Company, for the United States Department of Energy's National Nuclear Security Administration under Contract DE-AC04-94AL85000. This initial, proof-of-concept capability was funded through the Navy's (N-81) World Class Modeling program; and leveraged over thirty years of R&D funding at Sandia.

---

[1] Next stage efforts should certainly include a formal regimen of verification and validation testing.

# References

1. Ajzen, I., Madden, T.J. (1986). Prediction of goal-directed behavior: Attitudes, intentions, and perceived behavioral control. *Journal of Experimental Social Psychology, 22*, 453-474.
2. Backus, G.A. & Glass, R. (2006). *An agent-based model component to a framework for the analysis of terrorist-group dynamics.* Sandia National Laboratories, Technical Report: SAND2006-0860.
3. Bernard, M.L., Xavier, P., Wolfenbarger, P., Hart, D., & Waymire, R., Glickman, G. (2005). Psychologically plausible cognitive models for simulating interactive human behaviors. *Proceedings of the Human Factors and Ergonomics Society 49th Annual Meeting.* Orlando, FL.
4. Boslough, M., Sprigg, B.J., Backus, G.A., Taylor, M., McNamara, L., Fujii, J., Murphy, K., Malczynski, L., & Reinert, R. (2004). *Climate change effects on international stability: A white paper.* Sandia National Laboratories, Technical Report, SAND2004-5973.
5. Engle, R.F. &. Granger, C.W.J. (1987). Co-integration and error correction representation, estimation, and testing, *Econometric, 55*, 251-276.
6. Engle, R.F., & Granger, C.W.J. (1991). *Long-Run Economic Relationships: Readings in Cointegration,* Oxford University Press, Oxford, UK.
7. Fishbein, M., & Stasson, M. (1990). The role of desires, self-predictions, and perceived control in the prediction of training session attendance. *Journal of Applied Social Psychology, 20*, 173-198.
8. Forsythe, C. & Xavier, P. (2002). Human emulation: Progress toward realistic synthetic human agents. *Proceedings of the 11th Conference on Computer-Generated Forces and Behavior Representation,* Orlando, FL.257-266.
9. Gershenfeld, N.A. (1998). *The Nature of Mathematical Modeling.* Cambridge University Press.
10. Keeney, R.L., and Raiffa, H. (1976) *Decisions with Multiple Objectives.* John Wiley & Sons, New York, NY.
11. Hendry, D. F. (1993) *Econometrics: Alchemy or Science?* Blackwell Publishers, Cambridge, UK.
12. McNamara, L.A., et. al, (2008). *R&D for Computational Cognitive and Social Models: Foundations for Model Evaluation through Verification and Validation, Sandia National Laboratories,* Technical Report: SAND2008-6453.
13. McFadden, D. (1982). "Qualitative Response Models," in *Advances in Econometrics.* Ed. Werner Hildenbrand, Cambridge University Press, New York.
14. Sterman, J., (2000). *Business Dynamics: Systems Thinking and Modeling for a Complex World.* McGraw-Hill/Irwin, Boston.

# The Use of Agent-based Modeling in Projecting Risk Factors into the Future

Georgiy V. Bobashev[†], Robert J. Morris[†], William A. Zule[†], Andrei V. Borshchev[*],
Lee Hoffer[‡]

[†] RTI International, Research Triangle Park, NC
[*] XJTEK, St. Petersburg, Russia
[‡] Case Western Reserve University, Cleveland, OH

**Abstract** Human behavior is dynamic, which means that it changes and adapts. Health sciences, however, often consider static risk factors measured once in a cross-sectional survey. Population or group outcomes are then linked to these static risk factors. In this paper, we show how the use of agent-based models allow one to consider risks in a dynamic sense, i.e., to estimate how risk factors affect future outcomes through behavior. We illustrate the issue of dynamic risks using the examples of the heroin market and HIV transmission on sexual and drug-using networks. We show how the social hierarchy among drug users impacts the order of injection and thus the probability of HIV-free survival. We also illustrate the role of street brokers in the functioning of the heroin market. Although the results do not have the same validity as the data obtained from a longitudinal study, they often provide good insight into underlying social mechanisms without the need for conducting expensive and often unfeasible longitudinal studies.

## 1 Introduction

Simulation modeling can have numerous objectives. Some objectives are focused on predicting a number (and a confidence interval) corresponding to a future outcome, while others are focused on understanding mechanisms and the ways to influence them in order to change an outcome. These objectives often call for different modeling approaches. Predictive models are usually based on statistics and thus are deeply grounded in data. Models that are focused on understanding mechanisms might be data-free but should be grounded in theoretical reasoning. When uncertainties are applied in the same way to two or more competing interventions, qualitative analysis can indicate where in the parameter space one intervention has an advantage against another. This could be sufficient to help a policy maker evaluate the interventions.

H. Liu et al. (eds.), *Social Computing and Behavioral Modeling*,
DOI: 10.1007/978-1-4419-0056-2_8, © Springer Science + Business Media, LLC 2009

In order to project an outcome into the future, one usually needs the following:

- clear definitions of the outcome values, i.e., numeric or categorical values;
- initial values of the outcomes and risk factors, i.e., the values of the outcomes at the starting (initial) time point (e.g., at baseline [time zero] an individual could be either HIV positive or negative);
- descriptions of the actions that individuals may take in the future (e.g., having sex without a condom);
- the collection of factors determining the actions, which could relate to a number of independent variables as well as the past actions or past states simultaneously (e.g., having a history of heroin use makes a person more likely to use it in the future); and
- the translator of the behaviors into the outcomes (e.g., HIV is transmitted per direct syringe-sharing with a certain probability, which is lower if the syringe is rinsed in bleach).

Beyond understanding the objectives, a clear understanding of the scales on which the outcome resides determines the choice of modeling approach, such as a system dynamics, process, or agent-based model [1, 4]. Agent-based models have the advantage of being more flexible and accommodating of very complex behaviors. However, the cost of such flexibility is often intractability, i.e., the presence of too many parameters makes estimation too difficult.

Behavior studies bring specific challenges to modeling because behavior is adaptive and often cannot be described by a system of difference or differential equations. It is also difficult, expensive, and sometimes unfeasible to collect longitudinal data that would allow for direct and more precise measurement of model parameters and trajectories. Thus, we propose an approach that uses data based on the cross-sectional surveys or ethnographic data but allows for addressing qualitative and crude quantitative features of particular behavioral and health outcomes. We present two examples: One is the effect of social hierarchy on the risks of HIV, and the other is the role of street brokers in the functioning of heroin market.

## 2 Projecting HIV Factors into the Future

At the beginning of the HIV epidemic in the United States, researchers found that heroin users who had been injecting for a longer time had a lower prevalence of HIV than drug users with a shorter history of injecting. It turned out that a social hierarchy that had developed within heroin-using culture was responsible for this phenomenon. Those who had used heroin longer were likely to be higher on the social hierarchy of users and were more likely to procure heroin. Thus they were more likely to inject before the others, which means that an infected syringe would be less likely to reach them. Such an effect is difficult to model using system dynamics methods, but agent-based modeling allows us to incorporate such a modification easily. The model describes the community of drug users where few users are infected with HIV. As time passes, more drug injectors become HIV positive. In particular, the model setting was the following:

- Each injection drug user has a network of "buddies" (total n=750).
- Sharing occurs on a daily basis with a fraction of users (20%).
- Syringes are used in the order starting from most experienced to least experienced.
- Initial infected percentage is 10%.
- If a syringe passes through an infected person, it becomes infected.
- Probability of infection given the contact with an infected syringe is 0.0008.
- Individuals who die are replaced with the newly uninfected.
- The average survival after acquiring HIV is 5 years

Simulations were conducted over a 5-year period, and at each year and we outputted the time when each agent became HIV positive (if at all). We can compare the HIV status of individuals with long and short histories of injecting. By averaging over a number of replications and a group of subjects, we are able to collect the necessary statistics and estimate the difference in survival from HIV between those who had more than 2 years of drug-using experience and those who had less. Figure 1 indicates that the survival is indeed better among those who had used the drug for a longer period of time. We could use a variety of statistical methods to estimate odds ratios, relative risk, hazard, etc., by applying appropriate statistical methods to the simulated data in the same manner as it would be applied to the real data. Although the level of uncertainty is relatively high, we assured the credibility of the results by obtaining model parameters from peer-reviewed epidemiological studies. While there is very little data to validate the model independently, the results uncover the mechanism that explains the phenomenon observed by HIV ethnographers.

Figure 1. Survival curves among the highly experienced injectors (solid black line) and not very experienced (dashed red line) injectors.

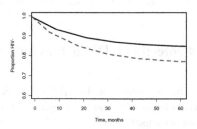

# 3 Projecting the Effects of Drug Busts on the Dynamics of the Heroin Market

Narrative and descriptive accounts collected by Hoffer's ethnographic work [2] were used to develop an agent-based model of the heroin market in Denver CO [3]. This ethnographic work determined the structure of the market, and defined agent roles and the rules according to which the market operated. In particular, the agents within the market were drug users, street dealers who purchased drug supplies from the private dealers and sold them to the users, street brokers who linked users with dealers (street and private) for a small portion of the purchased drug, and police and homeless people who happen to be present at the market location. When possible, behaviors, how activities were organized, decision-making processes, interactive activity, and relations between agents were programmed as described by participants. For example, private dealers described the process of using brokers to gain introductions to customers. Behaviors were also programmed based on observation of the participants.

The model contained about 500 users, 30 street dealers, and a varied number of street brokers. The market functions on a daily basis, and the amount of purchased heroin is determined by the individual agent's habit. Each day, an agent would go through a decision-making process where the decisions are based on a number of factors such as the cost of heroin, dealer availability, dealer ranking in terms of trust, presence of the police, etc.

The users would go to the market and try to locate the familiar dealer from the list of trusted dealers or would use the services of a street broker who could eventually make a direct contact with the private dealer. Police busts were simulated by the removal of the majority of street dealers. More details of the model are described in Hoffer and Bobashev [3].

We estimated the effect of the drug busts when police would arrest a number of street dealers on the market slowdown and recovery and compared the speed of the recovery with and without the street brokers.

At the end of each day, we collected information about the number of successful transactions, the amount of heroin purchased, and the types of transactions (i.e., with a private dealer, a street dealer, or a broker).

In Figure 2, we present curves (averaged over 100 replications) that describe the number of transactions in the market with and without street brokers and the speed of market recovery after two sequential police busts.

Figure 2. The number of drug transactions in the market with 0, 25, and 50 street brokers when police busts arrest most of the street dealers.

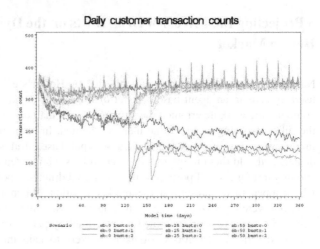

As seen in Figure 2, after two police busts, the market recovers more slowly and at a lower level of transactions than when the street brokers are present. In fact, after the busts, street brokers provide the users with better connections with the private dealers, which makes the market more efficient and support more transactions. The biweekly spikes in the number of transactions indicate that when many users get paid (assuming semi-monthly payment), they can purchase the drug immediately and not need to wait till the street dealer "restocks" after selling all the drugs. Thus the role of street brokers is very significant in sustaining the market after the bust, and this role if often overlooked in the description of drug markets.

# 4 Discussion

We have presented two examples of the use of the multi-agent models and showed how they could be useful for understanding and projecting risk factors into the future. These models uncover some of the relationships that should be considered when developing policy based on personal and community risk factors. Although the results are not "validated" in the sense of predictive models, they provide a state-of-the-art approach to quantify the critical elements of behavior and provide an important qualitative insight on potential mechanisms that until now had only been described verbally by ethnographers. Because the actual behavior and the outcomes could be very complex (detailed, nonlinear, evolutionary, and multiscale), the modes focusing on any specific area of research are likely to be of a much higher complexity level. However, in this presentation we would like to emphasize the role of agent-based modeling in providing insights that are not obvious without such models. The HIV example reconstructed historic evidence, and the heroin market example formalized new evidence that should be explored further. Both examples illustrated the importance of certain micro-scale behavior components in macro-level results. While such micro-components are difficult to capture through the use of equation-based models, agent-based modeling makes such descriptions feasible.

# References

1. Colizza V, Barthelemy M, Barrat A, Vespignani A (2007) Epidemic modeling in complex realities. CR Biologies 330:364-374
2. Hoffer L (2006) Junkie business: The evolution and operation of a heroin dealing network. Belmont, CA, Thompson Wadsworth
3. Hoffer L, Bobashev GV (in press, 2009) Researching a local heroin market as a complex adaptive system. American Journal of Community Psychology
4. Riley S (2007) Large-scale spatial-transmission models of infectious diseases. Science 316:1298-1301

# Development of an Integrated Sociological Modeling Framework (ISMF) to Model Social Systems

LeeRoy Bronner[†] and Babatunde Olubando[*]

[†] leeroy.bronner@morgan.edu, Morgan State University, Baltimore, MD
[*] tunde.olubando@gmail.com, Morgan State University, Baltimore, MD

**Abstract:** Social simulation uses computer based architecture to model and analyze social systems. Simply, a social system is the people in a society considered as a system organized by similar patterns of behavior. Object-Oriented Analysis and Design (OOAD) is a software engineering methodology that can be used to analyze and model social systems. Social systems can be very complex; how do we translate this complexity into the computer environment? Systems and software engineers are adept at deciphering complexities involved in the development of software systems using a variety of methodologies. Integrating software engineering development concepts with the research and development of social systems provides a way for the same disciplined approach used in producing zero-defect software to be used to improve sociological research methodology. This paper proposes the development of a framework that integrates software engineering development methodology with concepts of sociological research for studying social phenomena.

## 1 Introduction

### 1.1 Terms

Joint Application Development (JAD): a workshop centered on bringing a cross disciplinary functional teams together to tackle a problem or arrive at a decision.
Mind Map: a mind map is a diagram used to represent ideas, tasks or other items linked to and arranged radially around a central key concept which aids in the process of decision making.
Actor: an actor is a user that interacts with a system. The actor represents a role that a user can play in the development of a use case.

H. Liu et al. (eds.), *Social Computing and Behavioral Modeling*,
DOI: 10.1007/978-1-4419-0056-2_9, © Springer Science + Business Media, LLC 2009

Use Case: a use case (i.e., a system requirement) describes the sequence of steps a system performs to accomplish a task as it relates to a particular actor.

Object-Oriented Analysis and Design (OOAD): a methodology used in software development to develop software systems based on a collection of interacting objects.

Framework: a logical structure for classifying and organizing information.

Model: a general framework for the representation of a system that allows for reasoning about and investigation of the properties of the system.

Architecture: the specification of the parts and relationships among these parts of a system and the rules for the interactions within these relationships.

Enterprise Architect (EA): a comprehensive UML analysis and design tool, used for system development from requirements gathering, through to the analysis stages, design, testing and maintenance. Also, it is used to archive and manage model development artifacts.

## 1.2 Background

The traditional approach for performing social research often involves the use of a standard set of tools for determining the validity of the social phenomena being observed. These tools often enable the researcher to develop models or representations of the natural phenomena that explains the relationships between entities while discarding some other relationships [1, 2]. In some situations, these representations are in a purely narrative format and difficult to navigate and understand. Some of the issues that may arise from this narrative structure include inconsistencies, over-generalization of the relationships being examined, misinterpretation and large volumes of information. In some other areas of social research, mathematical and statistical models are used to explain social phenomena which may be confusing. These are some of the concepts that impact sociological research and form the basis for this research.

## 2 Methodology

## 2.1 Joint Application Development (JAD) Process

Joint Application Development is a popular technique used in the software engineering field that allows the potential users of a software system being developed to participate in the "fact-finding" process in conjunction with the developers. With the incorporation of mind mapping and use case methodology in the JAD session, it will empower the participants (i.e., sociologists and system analysts) in the session with a well-rounded, clear idea of how the system should perform. Through an iterative process, and constant feedback from all participants the output developed in the JAD session will be input to the next phase of the project which will entail the development of object-oriented models. In expanding the JAD session to focus on social systems, the session can be compared to the focus group that is used in social science research. JAD consists of five steps that enable users, stakeholders, and system analyst to arrive at an agreed-upon view of the system. These steps are: project definition, research, session preparation, the session, and production of the final JAD document [4]. At each step, there are required tasks to be performed to enable a successful implementation of the system.

As seen in figure 1, the first step of the JAD process, project definition, results in a Stakeholders Definition Guide which will aid the latter stages of the process. This guide contains the project scope, objectives, constraints, resource requirements, assumptions, open questions, and any processes that might be needed from a stakeholder perspective. The next step in the process is the Research phase. The Research phase involves the facilitator (i.e., system analyst) developing an understanding of the system. This entails creating models, gathering information about the processes involved

Figure 1: JAD Process with artifacts

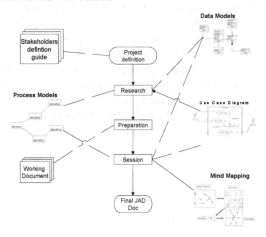

in each part of the system. As developed earlier in the project definition phase, interviews are essential for gathering information about the functions or processes involved in the system, if they exists already. The preparation phase of the JAD process entails the logistical setup for the actual JAD session. The working document is finalized and additional components are added, such as the session agenda.. A description of the requirements and other functional components of the system are detailed. A list of required session participants is generated and their attendance is confirmed for the meeting. All other issues such as distribution of the working document to participants before the session are resolved. Logistical issues relating to hosting the session are addressed in this phase of the process. Concerns about location of session, time required for the session, presentation aids, refreshments, training of the scribe and other concerns are resolved. The next phase of the JAD process is the actual session itself where complete system requirements are defined (i.e., use cases).

All models discussed in the session are translated into computer models using software development tools (i.e., Enterprise Architect, Rational Software Suite, etc.). With the development of mind mapping technology, some of the ideas generated during the interview can be formally gathered and included in the final working document of the JAD session. This document (figure 1) is the critical artifact that is generated from the JAD session and it contains all the data and process models developed in the session and other components such as use case models are added to the documentation.

## 2.2 Mind Mapping

Mind Mapping is a methodology developed on the theory of concept mapping [6]. It is based on the premise of conceptual pragmatism and cognitive economy, i.e. our ideas should have maximum use and minimum complexity. Mind maps allow JAD

Figure 2: Mind Maps at work

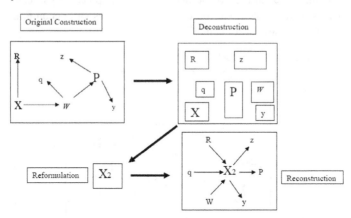

participants to reason about a central concept with branches to other related ideas that support a central idea. One example given in *Sheridan* is one where there is a construct of several ideas connected together. However the connections between the concepts are deemed vague and inconsistent. Mind Maps allow us to deconstruct and reformulate some of these ideas to enhance them and redevelop the overall construct. As shown in the figure 2, the theory behind mind mapping allows the user(s) to break down complex ideas into more meaningful thought processes. Based on this approach, mind maps can be used as the basis for carrying out JAD meetings. As can be seen in figure 2, the original model consists of several ideas that are connected together, although X is related to all sections of the model, it is not evident in the original model. However, through the process of mind mapping, a new model is developed to illustrate the linkage between all constructs of the model.

## 2.3 Use Case Modeling

Use case modeling is a technique used in the software development process where the developers model the functionality of the system from the user's perspective. Using use case technology, JAD session participants focus on the behavior of the system from an external point of view. Use case modeling allows the analyst to elicit all the major components and interactions involved in the system..

The use case model is made up of 3 components as shown in figure 3. They are the use case, use case diagrams and the use case scenarios. The use case scenario defines steps required to perform a certain task. Use cases are used to detail the basic functionality of the system. By detailing all the actors and their possible use cases, the complete requirements of the system can be defined. Use cases can become an effective communication tool for all participants [7].

Figure 3: Example Use Case Model

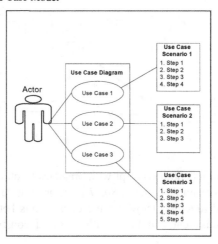

# 3 Enhanced Joint Application Development (E-JAD)

The enhanced joint application development (E-JAD) process is a modification of the JAD process. E-JAD incorporates use case modeling and mind mapping with the JAD paradigm. This new process develops a clearly defined analysis structure and enhanced modeling capabilities. Also, this enhanced process produces artifacts from the mind mapping paradigm to be encapsulated in a structured form that can be revisited during the stakeholder session phase of the JAD process.

When using all three processes individually, several artifacts are created independently. However with the integration of the processes, the use case model, mind maps as well as the final JAD document can be put together in one comprehensive document that covers all three areas of expertise for these processes.

Figure 4: Enhanced JAD Process

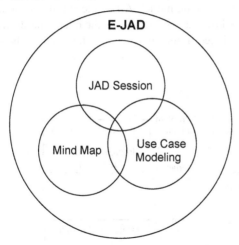

# 4 Object-Oriented Analysis and Design (OOAD)

The object-oriented design process is a subset of systems engineering, and can be used to analyze and model any type of system [8]. It functions on the principle of modeling systems as a set of collaborating objects. All objects contained in the system communicate with other objects through messages, and manipulation of object data. With the object-oriented approach of developing software systems, system and software engineers have been able to use the same set of models during the development process, thereby providing for a consistent development process.

Using the system development lifecycle process used in software development, the process starts with an initial problem definition. This involves the system users and developer in refining the models that will need to be used in solving the problem. Several iterations of this step are needed in most situations. In this step, several tasks are performed. This phase can be regarded as the requirements elicitation phase. However, the majority of this task is performed in the enhanced JAD phase of the process. The next phase of the OOAD system development lifecycle is the analysis step. It is primarily concerned with the development of abstractions which encompass the problem domain [3]. In producing a model for the problem, it is important that it is complete, verifiable and consistent with the problem domain being addressed. The analysis model also needs to be connected to the requirements gathered during the elicitation phase of the lifecycle. The analysis model is usually broken down into the functional model, the analysis object model and the dynamic model. All of the sub-models enhance the quality of the interactions modeled by the designer. Once the client and designer agree on the content and associations elicited in the analysis model, the client then signs off on the analysis model.

Figure 5: System Development Lifecycle

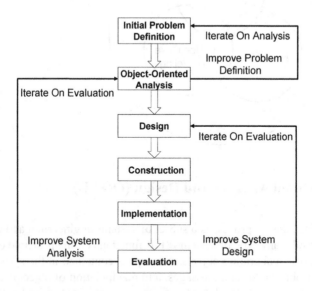

The next phase of the OOAD system lifecycle in which the analysis model is further refined in the design phase. It involves the implementation of the design goals in relationship to the systems requirements, the analysis model developed as well as the expansion and revision of the artifacts developed in the analysis phase of the OOAD lifecycle. This phase also involves some system decomposition in which the system is broken down into subsystems that address some of the system goals. In using systems engineering methodology in the development of social systems, the design phase of the OOAD lifecycle is where the similarities to the actual system development life cycle stop. The difference between the application of the lifecycle to social systems development is the implementation approach to the problem. After the design phase is completed, the artifacts produced are the proposed intervention to the social system. The implementation phase used is different in that that the proposed solution(s) to the social system is them taken out into the field and used as social interventions.

## 5 Integrated Sociological Modeling Framework (ISMF)

The Integrated Sociological Modeling Framework (ISMF) is an architecture that enables the social scientists and system analysts to arrive at a well defined structure for a social system. This structure will provide a general understanding of the parameters associated with the model to be developed. With all the components discussed in the methodology section of this paper, a seamless approach has been developed that integrates all these components in a framework that allows for improved sociological research and development studies. Using sociological methodology and software engineering technology, an interdisciplinary project can result in greater study benefits. As shown in the diagram below, social scientists and system analysts meet to discuss the social system to be analyzed or developed. The discussion will take place within a JAD session. However, before the JAD session convenes, there are several tasks that must be performed before hand.

During the JAD session, mind maps and use case models are developed. These sessions take place over a period until all participants sign off on the final document prepared from the JAD session(s). The documents, which include all mind maps and use cases, are saved in a database for easy retrieval and access for use in other phases of the social study. The next phase of the project entails the use of object-oriented design methods to develop a model that represents the discussions and artifacts generated in the JAD session. This model defines the problem solution, namely, intervention.

Figure 6: Integrated Sociological Modeling Framework (ISMF)

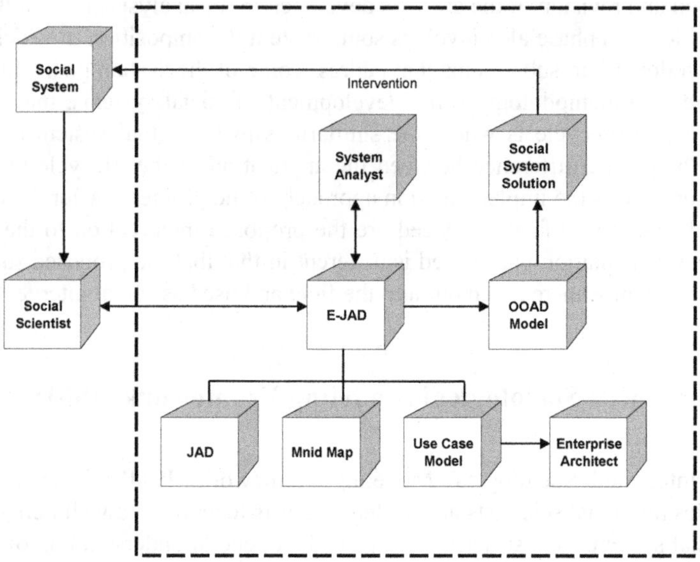

## 5.1 Example Social Study: Preconception Peer Education Program (PPPEP)

A research study involving a public health project is being used to test the feasibility of the ISMF. The study is being performed under the Office of Minority Health (OMH) in the U.S. Department of Health and Human Services (DHHS) in partnership with the Association of Maternal and Child Health Programs (AMCHP), CityMatCH and March of Dimes. This research addresses the problem of preconception by first implementing a training program entitled the *Preconception Peer Education Program* (PPEP) on a number of college campuses to impact the problem of infant mortality.

Using the ISMF framework, a model is being developed for the analysis and design of a plan for the implementation of the PPEP. The PPEP is divided into four phases as shown in figure 7. The phases are broken down from the initial training that occurred at the office of Minority Health to the last phase which entails community outreach to disseminate the study results.

Figure 7: PPEP Project Development Phases

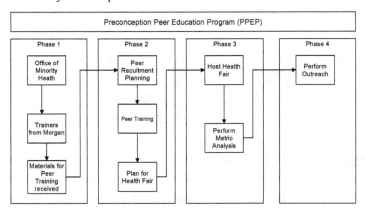

In using the ISMF framework, the facilitator, along with a system analyst, have started developing use case diagrams that depict the functions required to successfully complete the PPEP. These are shown in the figure 8. The use case models developed in collaboration with public health trainers allow for the documentation, archival and retrieval of the process used to accomplish the different phases of the PPEP. The use case model shown in figure 8 depicts the Recruitment Super Use Case which involves all the use cases needed to accomplish the recruitment task for the PPEP. Each use case includes a use case scenario; an example of one use case shown in Table 1[8] defines the Recruit Peers use case. With each use case and its associated template, the object-oriented development process further requires the development of additional model artifacts (i.e., class, object, sequence and activity diagrams) [7] to complete the model.

Figure 8: Recruitment Use Case Diagram (Drawn with EA)

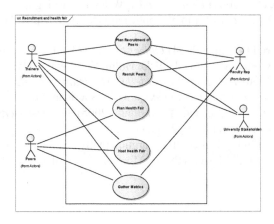

Table 1: Use Case Template

| Use Case Name | Recruit Peers |
|---|---|
| Use case ID | UC-101 |
| Super Use Case | Recruitment |
| Actor(s) | Trainers, Faculty Rep, Peers, University Stakeholders |
| Brief Description | This use case details the steps involved in recruiting peers for the PPEP project. |
| Preconditions | Trainers must be hired. |
| Post-conditions | Peer recruitment achieved |
| Flow of Events | Trainers meet with Faculty Rep to Discuss Department Involvement |
| | Trainers consult with University Stakeholders in selecting possible peers |
| | 3. Trainers send out peer invitation packets to selected peers |
| | 4. Peers respond to invitation from Trainers |
| | 5. Trainers and Faculty Rep hold kick off meeting to introduce PPEP |
| Priority | High |
| Issues | Not enough peers accepted invitation for PPEP involvement |

# References

1.  E. Zeggelink, et al (1996) "object-oriented modeling of social networks," *Computational & Mathematical Organizational Theory*, vol. 2, no. 2, pp. 115–138.
2.  N. Gilbert (2004) "Agent-based social simulation: dealing with complexity," Centre for Research on Social Simulation University of Surrey Guildford UK, Tech. Report.
3.  H.E. Eriksson, et al (2004) *UML 2 Toolkit*. Wiley Publishing Inc., 2004, ch. 1, p. 13.
4.  J. Wood and D. Silver (1995) Joint Application Development, T. Hudson, Ed. Wiley & Sons.
5.  a. A. H. D. Bernd Bruegge (2004) *Object-Oriented Software Engineering: Using UML, Patterns and Java*, 2nd ed., T. D. Holm, Ed. Prentice Hall.
6.  W. Sheridan (2006) *The Human Knowledge Mindmap: A guide to the mindset needed to perform competent knowledge work*, Sheridan, Ed.
7.  C. H. Tsang, C. S. Lau, and Y.K.Leung (2005) *Object-Oriented Technology*. McGraw Hill Education (Asia), 2005, ch. 1, p. 10.
8.  J.-D. Fletcher (2007) "Integrating argumentative rationale within an object-oriented framework," Master's thesis, Morgan State University, Baltimore, MD.

# The Coherence Model of Preference and Belief Formation

Sun-Ki Chai[*]

[*] sunki@hawaii.edu, University of Hawai'i, Honolulu HI

**Abstract:** Rational choice is the dominant approach in the social sciences to modeling individual and collective behavior. Recently, however, a great deal of criticism has been directed at it, much of from practicioners of the approach itself. Perhaps the strongest criticism has been directed at rational choice's inadequate modeling of preferences and beliefs. The coherence model presented here accounts for preferences and beliefs in a way that is applicable across the full range of contexts where conventional rational choice models can be applied and is compatible with assumptions of rational optimization. It will be based on the assumption that individuals will adjust preferences and beliefs to minimize expected regret, where expected regret is defined as the difference in perceived expected utility between each action taken and the action retrospectively viewed as optimal. Coherence is defined as attaining a state of zero expected regret is referred to as coherence, and is seen as the ultimate "meta-goal" of individuals engaging in "choosing" their preferences and beliefs with a specified set of reality constraints. This model designed to integrate a range of empirical findings and theories about the construction of self from a variety of social science disciplines.

## 1 Introduction

Rational choice is the dominant approach in the social sciences to modeling individual and collective behavior. Recently, however, a great deal of criticism has been directed at it, much of from practicioners of the approach itself. Perhaps the strongest criticism has been directed at rational choice's inadequate modeling of preferences and beliefs. While a large literature has arisen to examine alternative assumptions about preferences and beliefs, these new works typically limit themselves to a fairly narrow scope or leave unspecified parameters, creating the impression that one must trade away the strengths of the conventional approach in order to attain greater realism. However, this is not necessarily the case. The coherence model presented here accounts for preferences and beliefs in a way that is generalizable across the full range of contexts where conventional rational choice models can be applied and is compatible with assumptions of rational optimization. It will be based on the assumption that individuals will adjust preferences and beliefs to minimize expected regret, where expected regret is defined as the difference in perceived expected utility between each action taken and the action retrospectively viewed as optimal. Coherence is defined as attaining a state

H. Liu et al. (eds.), *Social Computing and Behavioral Modeling*,
DOI: 10.1007/978-1-4419-0056-2_10, © Springer Science + Business Media, LLC 2009

of zero expected regret is referred to as coherence, and is seen as the ultimate "meta-goal" of individuals engaging in "choosing" their preferences and beliefs with a specified set of reality constraints. This model designed to integrate a range of empirical findings and theories about the construction of self from a variety of social science disciplines.

The goal of the coherence model is to provide a general model of preference and belief formation, one that is consistence with a rational optimization model of action characteristic of the rational choice approach. It addresses a wide range of criticisms that have been made against the conventional rational choice approach, particularly the accusation that its view of preferences and beliefs is unrealistic and leads to inaccurate or indeterminate predictions. Actually, there are two versions of the rational choice approach that can be recognized. One is "thin" rationality, in which the nature of preferences and beliefs is left unspecified, and the other is "thick" rationality, in which preferences are seen as self-regarding, materialistic, and isomorphic across all individuals; and beliefs are seen as being completely determined by observation and logical inference from that observation (Chai 2001, 5-8). In practice the thick version is what is generally associated with conventional application of the rational choice approach, since the thinner version is not capable of making predictions on its own.

However, the conventional approach has been attacked from a number of directions as being unrealistic as a portrayal of human nature, as of limited use in making predictions outside a narrow realm of economic behavior. It is clear that one way to improve the predictive accuracy of the model would be to generate new models of preference and/or beliefs. Much of the recent activity has taken place in the growing fields behavioral economics, where the concept of *social preferences* has taken hold as an antidote to the conventional self-regarding preferences (Rabin 1993; Bolton and Ockenfels 2000, Fehr and Schmidt 1999; Andreoni and Miller 2002; Charness and Rabin 2002), but similar activity is also taking place among rational choice practitioners in political science and sociology (for review, see Chai 1997). However, these approaches purposely generally posit an alternative set of uniform preferences and do not attempt to predict the variations in preferences that exist between individuals and groups, and within individuals and groups over time.

The coherence model is aimed at addressing just these sorts of variations by presenting a dynamic model of individual preference and belief formation. Moreover, because the process through which preferences and beliefs are formed are taken as inherently social, involving both collective action and communication, the model is also intended to be one which can explain the generation of social identity and culture as collectively held attitudes.

## 2 Specifying the Model

*Regret* for a single action, given a particular state of the world, is the difference between the maximal utility possible in that state and the utility provided given the chosen action. *Expected regret* for a single action is the regret for each possible state of the world multiplied by the perceived probability of that state's occurrence.

The simplest example of how expected regret might be calculated is in a dichotomous choice set with alternative actions $a$ and $b$, and two possible states of nature $s_1$ and $s_2$, which are perceived to have probabilities $p_1$ and $p_2$. This creates possible outcomes $U_{a1}$, $U_{a2}$, $U_{b1}$ and $U_{b2}$, where $U_{a1}$ stands for the utility resulting from action $a$ in state of the world $s_1$, and so forth. The expected regret for choosing action $a$ would then be $d = p_1 (\max(U_{a1}, U_{b1}) - U_{a1}) + p_2 (\max(U_{a2}, U_{b2}) - U_{a2})$.

For more complex choice sets with choices $a_1 \ldots a_n$, possible states of nature $s_1 \ldots s_k$, perceived probabilities $p_1 \ldots p_k$ for each state, and where utilities $U_{ij}$ stands for the utility resulting from action $a_i$ given state of the world $s_j$, and and $U^*_j = \max(U_{1j}, \ldots U_{nj})$, expected regret for choosing action $a_1$ will be $\sum_{j=1}^{k} p_j (U^*_j - U_{1j})$. Where outcomes are continuously distributed along a single dimension and $\phi_s$ is the probability density function for $s$, this will be $d = \int_s (U^*_j - U_{1j}) \phi_s(s) ds$. It is clear from the definition that $d$ will always be non-negative. Based on this definition, expected regret can be calculated whenever individuals have beliefs about the probability distribution of the utilities for each actions in a choice set, hence can be calculated for any choice set over which the rational optimization assumption can be applied.

The total amount of expected regret an individual feels at any point in time will be her *cumulative expected regret*, the expected regret attached to all past actions. Hence cumulative expected regret calculated at period $t$ will be $D(t) = \sum_{i=0}^{t} \delta^{t-i} d(i)$, where $0 < \delta \le 1$ is some optional geometrical discounting factor. Finally, *coherence* will be the state of having cumulative expected regret equal to zero.

The key assumption of this model will be that individuals make mental adjustments to minimize their level of expected regret:

**Assumption 1**: At each period, individuals will adjust their preferences and/or beliefs in order to allow for minimal cumulative expected regret over their entire sequence of past and prospective actions and choice sets.

In other words, individuals will attempt to attain a state of coherence by locating a set of preferences and beliefs for which the optimal set of actions available provides them with the lowest amount of expected regret. A state of coherence, if feasible, allows an individual to believe with certainty that she can calculate the very best plan for maximizing a destiny in life that she has chosen. This in turn allows the individual to view her life in a teleological manner, with all her actions moving her closer to this ultimate destiny.

If we leave things at that, this kind of "meta-optimization" would present difficulties. There could be degenerate ways of achieving coherence, such as being indifferent over all outcomes or adopting beliefs totally at odds with the observed world. Moreover, the search space for minimizing expected regret would be infinitely large, with any conceivable utility function or set of beliefs in play as candidates for the individual's sense of self. This in turn leads to a number of other assumptions placing constraints on beliefs.

**Assumption 2**: Individuals beliefs cannot exclude facts that derive from direct observation or logical inference.

This is in a sense a weaker form of the "information assumption" found in conventional rational choice theories, whereby all beliefs must derive from direct observation or logical inference. If this limit did not exist, individuals could achieve coherence by living in fantasy worlds where expected regret-inducing aspects of the environment simply disappear.

**Assumption 3**: Each individual's utility function will be constrained so to have a linear relationship of fixed slope with personal material welfare.

This assumption normalizes utility functions so that individuals cannot eliminate expected regret completely by adopting a "yogic" utility function in which all outcomes have equal utility. Such a utility function, needless to say, would make predicting action impossible. The assumption furthermore ensures that expected regret-reduction cannot occur simply by dividing the utility function by an arbitrarily large number.

**Assumption 4**: The parametric form of individual utility functions and beliefs are limited to the sum or conjunction of those forms that are available from ideologies found in the individual's cultural environment. However, any coefficients in an ideology can be shifted or negated arbitrarily by and individual in the process of preference and belief adjustment..

This may need a bit of more explanation. Basically, what it is positing is that no variable or functional form of relationship between variables can appear in a belief or utility function unless there is some explicit ideology that is the individual has been exposed to that states that this variable or relationship between variables is significant.

For instance, if an individual is to have a utility function $U = \ldots + x + \ldots$, then there must be an ideology that states that the good $x$ is a useful thing to have. However, once the ideology makes $x$ available to the utility function of an individual, the individual can choose to disvalue the good, i.e. $-x$ and/or place any weight upon it she wishes, i.e. *2x*.

**Assumption 5**: The process of minimizing expected regret will involve considering first the variables and variable functions found within ideologies in a sequence based on the salience of the ideologies in the cultural environment, where salience is a matter of the frequency with which an individual is exposed to communications espousing that ideology.

What this means is that that those ideological messages that are broadcast more often at an individual, either through the mass media or through personal contacts, will have precedence in shaping that individual's preferences and beliefs. This does not mean, of course, that individuals will passively absorb such ideologies. An ideology may be rejected outright (i.e. not incorporated in preferences and beliefs) if its incorporation does not lead to reduction in expected regret. Likewise, even if it is incorporated, manipulation of coefficients may lead to an opposite affective disposition towards an outcome or belief in an opposite causal relation than that espoused by the ideology.

This model is not at all meant to be a purely formal construction, but rather to adopt and integrate concepts and findings from a number of different literatures in multiple social science disciplines. Quite briefly, the formal definition of regret used here in versions of regret theory that have been proposed in economics (Savage 1972; Loomes and Sugden 1982; 1987; Bell 1982), the main difference being that the earlier versions have used regret as a model of decision-making, rather than preference and belief formation. Indeed, in its substantive implications, the model is closer to psychological theories of dissonance and their close cognates (Aronson 1992; Harmon-Jones 1999; Cooper 2002). While dissonance has been defined in varying and contradictory ways, newer versions of it in particular focus on the importance of action that seems to go in some way against an individual's own priorities or values. Likewise, the model is in line with recent sociological thinking on the construction of identity, particularly literature which focuses on the role of narrative and memory in creating a sense of collective purpose out of the past (Ricoeur 1988; 1992; Zerubavel 2004; Olick 2003; 2007). A number of other ties can be made, including narrative psychology (McAdams et al. 2006) and symbolic interactionism (Denzin 2007).

# 3 Implications for Preference and Belief Change

To begin, the focus will be on preference change that consists of increasing or decreasing weights on variables in linear utility functions. In other words, given a utility function in the form $\alpha_1 x_1 + \alpha_2 x_2 + \ldots + \alpha_n x_n$, where $x$'s represent variables and $\alpha$'s represent their weights, the focus will be on how shifts in $\alpha$'s affect expected regret.

**Proposition 1**: An individual can reduce expected regret for a past or intended action by raising the utility coefficient of a variable that is believed to be positively linked causally to the action.

Let the variable $x$ stand for a good or outcome which, by prior belief, is thought to be causally linked to the chosen action $a_1$. Without loss of generality, we can assume $x=1$ if the action is chosen and $x=0$ if not (this simply represents a choice of measurement units). The most natural interpretation of $x$ would be the intrinsic value placed upon the action itself or its direct consequences. In this case, increasing the weighting of $x$ by some amount $\Delta\alpha$ will shift the distribution for $U_{1j}$ - $U_{ij}$ to the right by $\Delta\alpha$ units for all actions $i \neq 1$, in which case $d^* = -\int_s \max(U^*_j - U + \Delta\alpha, 0) \, \phi_s(s) \, ds$. It is straightforward from then on to show that this will reduce expected regret compared to $d^* = -\int_s \max(U^*_j - U, 0) \, \phi_s(s) \, ds$ if there is at least some state of nature with positive probability density such that $U^*_j - U > 0$.

This theorem conforms with psychological findings on commitment, which reveal that individuals tend to increase their relative liking for alternatives after they have decided upon them (Brehm 1956). Commitment phenomena are often applied to explain the effectiveness of such influence techniques as "low-balling," where individuals are provided incentives to make choices, have some of their original incentives unexpectedly withdrawn, yet stick to their original choices because of post-decision increases in their intrinsic valuation (Cialdini, 1986). Finally, the theorem can be used to explain the "functional autonomy of motives," whereby means chosen to achieve particular ends are eventually valued intrinsically.

It also sheds light on the contradictory mechanisms Elster calls "sour grapes" and "forbidden fruit is sweet" (Elster 1983, chap. III). The first occurs when individuals adjust their preferences to adapt to their perceived circumstances (i.e to increase their expected utility), while the second occurs when adjustment is in the opposite direction. While Elster does not attempt to specify when these each of these opposing mechanisms will occur (and indeed believes it is impossible to do so), the theorem provides some indication that the effect will depend on the extent to which an individual has chosen to challenge the status quo in pursuit of new set of circumstances. Acceptance of the status quo breeds sour grapes for those new circumstances, while an act or intention to challenge will generate forbidden fruit.

As we move to beliefs, we can define the *net provision* of a good for an action as the amount of the good provided by that action compared to the highest amount provided by all actions available in the corresponding choice set. In this section, the focus will be on belief change that involves changing the expected value and the perceived variation of in the net provision of a particular variable for a chosen action.

**Proposition 2**: An individual can reduce expected regret for an action by increasing the expected net provision of a variable with positive utility weighting.

Let $y$ stand for some good or outcome with positive utility weighting $\alpha > 0$. Suppose the individual in question changes her belief set so that the distribution of $\hat{y}_{1j}$ shifts over $\Delta\hat{y}_1$ to the right for all states of nature $j$. Then, assuming the individual has no other beliefs which force her to simultaneously change the perceived distribution of other variables, the perceived distribution for $U_{1j}$ - $U_{ij}$ for all actions $i \neq 1$ is shifted $\alpha\Delta$ $\hat{y}_1$ to the right. The rest of the proof of theorem proceeds similarly to that for Theorem 1. Similarly, it can be shown that it is expected regret-reducing to reduce expectations of the net provision of a good with negative utility weighting.

72

This kind of belief adjustment corresponds to post-decision "bolstering", in which individuals adopt beliefs that emphasize the expected benefits that will the choice they have made, vis-a-vis the choice(s) that they have passed up (Knox and Inkster 1986; Younger, Walker, and Arrowood 1977). This can also be used to explain the "illusion of control" effect, where individuals have been found to systematically exaggerate the influence their choices have over important outcomes (Langer 1975).

These types of effects are relevant for the "wishful thinking" and "unwishfulthinking" phenomena in which individuals adjust their beliefs either towards or against expection of favorable outcomes (Elster 1983, chap. IV). The above theorem can specify the conditions under which each will occur. For an individual who has taken risks that depend on favorable circumstances occurring, expected regret can be reduced by wishful thinking, while for individuals who play it safe expected regret can be reduced by unwishful thinking.

# 4 Conclusion

This chapter has focused on simple processes of individual preference and belief change, but the coherence model can be used to explain both more complex processes of change at the individual level, such as the effect of exogenous structural changes on the extent of preference and belief formations, as well as collective processes of (for more detailed discussion, see Chai 2001, chap. 3). By applying these kinds of ideas it is possible to make predictions on such disparate outcomes such as the choice of economic (Chai 1998) and defense (Chai 1997a) ideology, as well as the formation of ethnic groups, their boundaries, and their propensity for violent collective action (Chai 1996; 2005). For ideology, the role of political activists, and particularly "ideological entrepreneurs" (Popkin 1979; Taylor 1989), is crucial in broadcasting messages that determine the alternatives that members of a population will consider first in altering their preference and beliefs in attempts to regain their coherence during times of rapid structural change. Those ideologues who propose preferences and belief changes that address the expected regret problems of the population will be most successful in getting their ideas adopted. For ethnic groups, the most important factors are the processes by which groups of individuals develop a collective identity and begin to mobilize on the basis of that identity. This is a multifaceted process that involves the acquisition of altruistic preferences (Collard 1978; Margolis 1982) and common social norms (Hechter and Opp 2005).

Overall, the coherence model is designed to provide an alternative to the conventional rational choice models that incorporates cultural, ideology, and identity into preferences and beliefs, but does so in a way that retains the generality, determinacy, and parsimony of the original model. Of course, there are bound to be multiple, and undoubtably superior, models that can fulfill these criteria, but it is important for the long-term development of rational choice analysis that more attempts be made in this direction.

# References

1. Allport, Gordon W. (1937) The Functional Autonomy of Motives. American Journal of Psychology 50:141-56.
2. Andreoni, James and Miller, John (2002) Giving according to GARP: An Experimental Test of the Rationality of Altruism. Econometrica 70(2): 737-53.
3. Aronson, Elliot (1992) The Return of the Repressed: Dissonance Theory Makes a Comeback (Special Issue). Psychological Inquiry 3(4):303-11.
4. Bell, David E. (1982) Regret in Decision-making under Uncertainty. Operations Research 33:961-81.
5. Bolton, Gary E. and Ockenfels, Axel (2000) A theory of equity, reciprocity and competition. American Economic Review 100:166-93.
6. Brehm, Jack W. (1956) Post-decision Changes in Desirability of Alternatives. Journal of Abnormal and Social Psychology 52:348-89.
7. Chai, Sun-Ki (1996) A Theory of Ethnic Group Boundaries. Nations and Nationalism 2(2):281-307.
8. Chai, Sun-Ki (1997) Rational Choice and Culture: Clashing Perspectives or Complementary Modes of Analysis. In Richard Ellis and Michael Thompson, eds,. Culture Matters. Boulder, CO: Westview Press.
9. Chai, Sun-Ki (1997a) Entrenching the Yoshida Defense Doctrine: Three Techniques for Institutionalization. International Organization 51(3):389-412.
10. Chai, Sun-Ki (1998) Endogenous Ideology Formation and Economic Policy in Former Colonies. Economic Development and Cultural Change 46(2):263-90.
11. Chai, Sun-Ki (2001) Choosing an Identity: A General Model of Preference and Belief Formation. Ann Arbor: University of Michigan Press.
12. Chai, Sun-Ki (2005) Predicting Ethnic Boundaries. European Sociological Review 21(4):375-391.
13. Charness, Gary and Matthew Rabin (2002) Understanding Social Preferences with Simple Tests. Quarterly Journal of Economics, 117:817-869
14. Cialdini, Robert B. (1986) Influence: Science and Practice. Glenville, Ill: Scott, Foresman.
15. Collard, David (1978) Altruism and Economy: A Study in Non-Selfish Economics. New York: Oxford University Press.
16. Cooper, Joel (2007) Cognitive Dissonance: 50 Years of a Classic Theory. Newbury Park, CA: Sage Publications.
17. Cote, James E. and Charles G. Levine (2002) Identity, Formation, Agency, and Culture: A Social Psychological Synthesis. New York: Lawrence Erlbaum.
18. Denzin, Norman K. (2007) Symbolic Interactionism and Cultural Studies: The Politics of Interpretation. New York: Wiley-Blackwell.
19. Elster, John (1983) Sour Grapes: Utilitarianism and the Genesis of Wants. Cambridge: Cambridge University Press.
20. Fehr, Ernst, and Klaus M. Schmidt (1999) A Theory of Fairness, Competition, and Cooperation. Quarterly Journal of Economics 114(3): 817-68.
21. Harmon-Jones, Eddie (1999) Cognitive Dissonance: Progress on a Pivotal Theory in Social Psychology. New York: American Psychological Association.

22. Hechter, Michael and Karl-Dieter Opp, eds. (2005) Social Norms. New York: Russell Sage Foundation.
23. Knox R.E. and J.A. Inkster (1968) Postdecisional Dissonance at Post Time. Journal of Personality and Social Psychology 8:319-323.
24. Langer, Ellen J. (1975) The Illusion of Control, Journal of Personality and Social Psychology 32:311-28.
25. Loomes, Graham and Robert Sugden (1982) Regret Theory: An Alternative Theory of Rational Choice Under Uncertainty. Economic Journal 92:805-24.
26. Loomes, Graham and Robert Sugden (1987) Some Implications of a More General Form of Regret Theory. Journal of Economic Theory 41(2):270-87.
27. Margolis, Howard (1982) Selfishness, Altruism and Rationality: A Theory of Social Choice. Cambridge: Cambridge University Press.
28. McAdams, Dan P., Ruthellen Josselson, and Amia Lieblich, eds. (2006) Identity And Story: Creating Self in Narrative. New York: American Psychological Association.
29. Olick, Jeffrey K. (2003) States of Memory: Continuities, Conflicts, and Transformations in National Retrospection. Durham, NC: Duke University Press.
30. Olick, Jeffrey K. (2007) The Politics of Regret: On Collective Memory and Historical Responsibility. New York: Routledge.
31. Popkin, Samuel (1979) The Rational Peasant: The Political Economy of Rural Society in Vietnam. Berkeley: University of California Press.
32. Rabin, Matthew (1993) Incorporating Fairness into Game Theory and Economics. American Economic Review 83(5):1281-1302.
33. Ricoeur, Paul (1988) Time and Narrative (vols. 1-3). Chicago: University of Chicago Press.
34. Ricoeur, Paul (2006) Memory, History, Forgetting. Chicago: University Of Chicago Press/
35. Savage, Leonard S. (1972) Foundations of Statistics, 2nd revised ed. New York: Dover Publishing.
36. Taylor, Michael (1987) The Possibility of Cooperation. Cambridge: Cambridge University Press/
37. Younger, J.C., L. Walker and A.J. Arrowood (1977) Postdecision Dissonance at the Fair, Personality and Social Psychology Bulletin 3:284-287.
38. Zerubavel, Eviatar (2004) Time Maps: Collective Memory and the Social Shape of the Past. Chicago: University Of Chicago Press.

# Cognitive Modeling of Household Economic Behaviors during Extreme Events

Mark A. Ehlen,[†] Michael L. Bernard,[*] and Andrew J. Scholand[††]

[†] maehlen@sandia.gov, Sandia National Laboratories, Albuquerque, NM
[*] mlberna@sandia.gov, Sandia National Laboratories, Albuquerque, NM
[††] ajschol@sandia.gov, Sandia National Laboratories, Albuquerque, NM

**Abstract** Traditional economic models of household behavior are generally not well suited for modeling the economic impacts of extreme events, due to (1) their assumptions of perfect rationality and perfect information; (2) their omission of important non-market factors on behavior; and (3) their omission of the unusual, scenario-specific conditions that extreme events pose on decision making. To overcome these shortcomings, we developed a cognitive-economic model of household behavior that captures and integrates many of these important psychological, non-market, and extreme-event effects. This model of household behavior was used in prototype simulations of how a pandemic influenza can impact the demand for food in a large metropolitan city. The simulations suggest that the impacts to food demand caused by household stress, fear, hoarding, and observing others doing the same could be far greater than those caused simply by the disease itself.

## 1 Introduction

Economic simulation tools give analysts the ability to assess the economic consequences of "extreme" events such as natural and man-made disasters. These simulations, however, generally use simplistic of "typical" economic behaviors that can be wholly unrealistic for extreme events. There are at least three causes of this inadequacy. First, these models often assume that individual economic actors (households, firms) have perfect knowledge of current and future conditions, of the future behaviors of each other, and of supply, demand, and prices in markets, and that they act rationally in both normal and extreme conditions. Specifically, the models do not capture essential information disparities and the lack of rationality that greatly influence "real-world" behaviors. Second, these models specifically do not include emotions, stress, and other potentially irrational psychological factors that greatly influence household economic decision-making. Finally, these models assume that the economic decision-making processes during an extreme event are the same as those during normal conditions, when in fact the informational, psychological, and cognition conditions are be very different.

H. Liu et al. (eds.), *Social Computing and Behavioral Modeling*,
DOI: 10.1007/978-1-4419-0056-2_11, © Springer Science + Business Media, LLC 2009

To explore and start to understand how these factors impact extreme-event economic behaviors and their aggregate effects on the economy, we developed a cognitive-economic, semantic memory network model of individual-household decision making that takes into account economic and environmental cues (or the lack thereof), psychology- and economics-based factors, and different processes of thinking based on the particular psychological condition of a household. This model was then implemented in large-scale agent-based simulations of a pandemic influenza to see how cognitive-based economic decision-making uses and influences the spread of disease; the spread of information about household conditions; the psychological health of households, neighborhoods, and society writ large; the economic behavior of households; and ultimately the adequacy of aggregate food supply.

This paper describes the cognitive-economic model, the large-scale agent-based economic model, and the primary findings from the pandemic simulations. Specifically, Section 2 outlines the structure, use, and validation of the cognitive-economic model of individual-household behavior. Section 3 outlines the agent-based model of household food demand during a pandemic influenza and describes general findings from these simulations. Section 4 summarizes and concludes.

## 2 The Cognitive-Economic Model[1]

The cognitive-economic model is a semantic memory network composed of (1) external *cues* that act as "inputs" to the overall cognitive decision process, (2) internal *concepts* that quantify a set of internal propensities or desires that result from interpreting the cues, and external *action intentions* or decisions that result from interpreting the balanced effects of the internal concepts. The cues are connected to the concepts, and the concepts are connected to the action intentions, via directed, weighted links that represent cognitive "signals" that convey the relative importance of each cue to the contexts and then the relative importance of each contexts to the action intentions.

Based on a review of work by others, the most salient set of external cues, contexts, action intentions, and links that capture household behavior during an extreme event is a set that focus specifically on factors related to (1) the specific scenario in question (in our case, a pandemic influenza), (2) key household psychological conditions that could influence decision making (stress, fear), and (3) the ultimate economic decision itself, in this case, how much food to purchase and hold. For the scenario herein, our set of external cues (each of which is valued as "yes" or "no") for an individual household includes: whether it sees other households getting sick with influenza, whether

---

[1] For more details on this model, see Kay et al (2007) and Smith et al (2007).

members of its own household get sick, whether it sees other households stockpiling food, whether it currently has enough food stored, whether federal authorities have asked it to stay home, whether it has enough funds to buy the amount of food it desires, and whether it sees a severe reduction of food at its local grocery store.

These cues then influence the relative importance of contexts that represent the following psychological priorities (each of which is quantified as an activation level of the context): the belief that the pandemic is in my community, the belief that the pandemic is spreading quickly, the belief that there is a high probability of family members getting sick, the belief that there is a moderate probability of family members getting sick, the desire to reduce "death anxiety," the desire to vaccinate family members, the willingness to not go to work or school, the desire to go to work due to lack of funds or due to work commitments, the willingness to severely reduce the number of trips to the grocery store, the desire to purchase more food, and the desire to stockpile food.

Finally, these contexts influence the relative importance of action intentions (each of which is valued as a number, and where many of these action intentions can be taken simultaneously and others are alternates to each other): stay at home, i.e., socially isolate; take flu preventative measures; go to work (or not); buy enough goods for one day; buy enough goods for two weeks, i.e., stockpile food; and do not buy goods today.

A key feature of the model is a set of nodes in the semantic memory network that activate two alternative emotional states, "high fear" and "low fear," which model the structural differences between the psychological states of normal and extreme-event conditions. These emotional-state nodes operate just like the context nodes in the network, but act as multiplexers: if a set of cues put the household in a state of "low fear," then it will use one set of low-fear context-to-action intention links; alternately, if the set of cues put the household in a state of "high-fear" (e.g., it gets infected and it sees other household stockpiling), it uses a set of high-fear context-to-action intention links (e.g., it desires to stockpile food). Also, because of known significant differences in extreme-event behaviors between people of high and low socio-economic status, caused by differences in economic resources, education, and availability of transportation, there are actually two different cognitive-economic semantic memory networks, a High Socio-Economic Status (HSES) network and a Low Socio-Economic Status (LSES) network. The two models have slightly different structures of nodes and edges, and use different values of edge weights.

Each of the salient model cues, contexts, and action intentions is based on psychology and pandemic research by others. The overall behavior of the cognitive-economic model was validated by testing all 128 "yes/no" combinations of the external cues to see if the action intentions that result from each combination make sense. For example, the cognitive model indicates that stress-related emotional states influence the propensity to get sick, that is, that a highly stressed household will have a greater chance of becoming infected. This mirrors the body of research that suggests that individuals under high stress are more likely to become infected due to psychosomatic influences on the human immune system.

## 3.0 Large-Scale Behavioral Modeling of Food Purchasing During a Pandemic Influenza

To understand how individual households that use this cognitive-economic decision making process interact with other households infectiously, socially, and economically, we implemented the model in the NISAC Agent-Based Laboratory for Economics™ (N-ABLE™), a large-scale agent-based simulation environment that models and simulates the impacts to and between firms and households. As a "laboratory," it provides an economic modeling platform from which new, important extensions of economic theory can be implemented, tested, and eventually applied to economics problems. It uses a data-driven, microeconomic model of individual private sector entities (private firms and households) to capture how changes in external conditions (largely infrastructure-related) impact and change the internal operations of and impacts to these private sector entities.

The standard N-ABLE™ model[2] of a household has a unit that produces "utility," the most common microeconomic measure of individual or household performance. To do so, it consumes one or more commodities from its warehouse or "cupboards." To fill its cupboards, it purchases these commodities from different supplying firms (e.g., retail supermarkets), based on availability and price at each firm. Firms that produce and supply commodities use fixed amounts of inputs, drawn from a warehouse and acquired by its buyers (or alternately, acquired "out of ground"), to produce a "good" that is stored in its warehouse and sold in markets. A food supply chain is created by vertically linking individual firms to one another and ultimately to households via commodity markets (e.g. manufactured food, wholesale, and retail) and infrastructure (e.g., roads, railroad, pipelines).[3]

To implement the cognitive-economic model in N-ABLE™, the utility-generating component of a household was modified to include the semantic memory network that can process the cues, contexts, and then action intentions. Pertaining to food demand, the cognitive model modifies the required level of food in the warehouse based on the model's desired number of days of food, which then changes the food buyer's desired purchase levels. The utility-generating component also includes a model of epidemiological disease spread so that a household can (1) contract disease through contact in

---

[2] For details on the N-ABLE™ model structure, ssee Ehlen and Eidson (2004).

[3] For examples, see Ehlen et al. (October 2007) and Ehlen et al. (March 2007).

public places (primarily schools and work); (2) go through contagious, asymptomatic, symptomatic, and infecting phases; and (3) ultimately either get well or die.

The resulting N-ABLE™ cognitive-economic household utility production function differs from the traditional economic utility function in several important ways: first, it assumes that the existence of family and community illness, public isolation, and food stockpiling may jeopardize future consumption, and thus allows the household to increase its local inventory of food to prevent future household food shortages. Second, the level of food inventories that a household currently has can influence (albeit indirectly) its decisions of whether to go to work, whether to isolate from others, and whether to take preventive health measures (e.g., take vaccines).

The primary means of validating the N-ABLE™ cognitive-economic simulations was comparison to relevant material generated by subject-matter experts. For example, these experts forecast that in a real pandemic, stress levels are likely to be very high: people will experienced significant distress due to loss of family members and anxiety about work, food, transportation, and basic infrastructure; public mood and social disruption and hoarding will be strong catalysts for problems. Households will try to isolate and self-medicate; and there would be a significant increase in sales at supermarkets, and consumers would resist leaving their homes. Finally, there are also likely to be supply chain-based shortages and hoarding will occur.[4]

To explore the use of the cognitive-economic model in an extreme event, we constructed a large-scale simulation of a pandemic influenza in Albuquerque, New Mexico. From a modeling perspective, this metropolitan city is ideal in that it is large, relatively isolated from other large cities (so that food supply and disease interactions are "local" and not influenced significantly by other cities), and has sufficient quantitative and geographic variations in socioeconomic status, infection circles, and food supply to see and test the relative importance of the different networks of causal influence (infection, information/cognition, and economic).

The N-ABLE™ model of this food economy is composed of approximately 250,000 households (source of data: City of Albuquerque [2008]) and the top 250 supermarkets in the metropolitan area (source of data: The D&B Corporation [2007]). Food supply from these supermarkets was calibrated to the average consumption of households, and a food market was created where all households have access to all supermarkets, but will go first to the supermarkets in their neighborhood until they run out of food, at which point they start shopping at next-nearest supermarkets.

Prior pandemic work has found that schools are a primary means of transmission of a virus between households (workplaces being second). To capture this "vector" of disease propagation, we created infection networks for each household based on each household's likely mix of children and the associated public schools they would at-

---

[4] For details on these validation conditions, see CSCFSC (2007), Food Marketing Institute (2006), Florida (2008), and White House (2005).

tend. To track the regional, neighborhood statistics on infection, emotion/cognition, and economics, statistics on the households were aggregated at the Census-tract and citywide levels.

Each household starts out healthy; over the course of the simulation a "Patient 0" is introduced as the first (and only) person infected with the disease. Each person who is or becomes infected has a constant probability of infecting others; this probability is typically called "R0," and is *a priori* unknown for a given strain of pandemic virus. Patient 0, due to interactions in the school and work networks, can (and does) infect others, causing the disease to spread. A fraction of these sick people die and are then removed from the simulation.

A series of simulations were conducted, following a specific set of categories designed to test the validity and usefulness of the model. Most of the simulations were of a length of 60 simulation days, where "Patient 0," or the first person infected with disease, was introduced on day 30 of the simulation. For the particular disease infection rate chosen and structure of infection networks, the disease spread and died out over about a 15-day period.

Fundamentally, the core of a large-scale simulation of a pandemic is how each individual household interprets and acts upon the change in information occurring outside and inside the household. Each household experiences and carries forward the following somewhat typical events and actions: first, before onset of the pandemic, the household is maintaining food inventories of 3 to 4 day-units (the amount of food the household eats in a day), well above its desired level of 1 day-unit. The fluctuations in daily inventory are caused by the alternating dynamics of daily consumption and going to the store for more food. At the onset of the pandemic, the Patient 0 household becomes sick. The cognitive-economic model activates the household into a high-fear state, causing it to stay at home and stockpile food. Given that its desired inventory of food is more than its current inventory, it runs to the store to get its desired level of food. Through social interactions, this household infects other households at a rate of R0.

On the order of 4-5 days after onset, other households begin to run out of food and get infected. They got to and maintain a high-fear, no-food, infected stage until they either get well or die. Later, these households that get well and finally acquire food, purchasing lower amounts than previously desired due to the fact that they are no longer ill. Given that they now has "enough food" and are no longer ill, they drop to the low-fear state, which has the combined effect of also lowering the desired inventory level back to normal. Unfortunately, on the order of 20-25 days after onset, many of these household, while no longer desiring to stockpile for their own reasons, witness others stockpiling, thereby causing them to re-enter a high-fear state again, again desire higher inventories of food, again stockpile, and thereby sending a signal for other, non-stockpiling households to stockpile.

Based on running simulations across a wide array of R0 values, structures of infection circles (school and work), and socio-economic conditions, our primary finding is that the food economy is very sensitive to households' desires to stockpile, whether it is caused by in-household illness, moving emotionally to a "high-fear" state, witnessing others stockpiling, or experiencing food shortages at the supermarket. While it is true that lower levels of R0 reduced the level of disease (and very low levels of R0 would put into question whether it is a widespread pandemic or extreme event at all), the invariant result is that (1) stockpiling and excess demand on the food network occurs often, and (2) the asynchronous nature of when and how much people purchase food causes the high probability of repeated periods of stockpiling well after the passing of the disease itself.

Public mood (e.g., panicking) and attendant behavior (e.g., social disruption, hoarding) are a significant catalyst for problems in our simulations. The acts of infection, staying at home, and witnessing hoarding of food by others raises individual-household stress levels to the point that hoarding of food becomes "viral" and widespread. As simulated, this hoarding significantly disrupts the functionality of the food supply chain, through severe demand shocks and supply shortages.

Our food economy is challenged by "panic stories," surge buying by consumer, and the inability of the supply chain to service demand. In our simulations, "panic stories" are the witnessing of households hoarding food for essentially unknown reasons (that is, it is not known whether a household is hoarding because it is sick, it is inherently panicking, or because it is witnessing hoarding itself). The food chain cannot inherently satisfy a viral surge in food demand of this size, due to the just-in-time nature of this supply chain, thereby causing further aberrant emotional, cognitive, and economic behavior.

Finally, prior N-ABLE™-based pandemic analysis (Ehlen et al. [October 2007]) shows that the U.S. manufactured food supply chain is inherently vulnerable to loss of food at localities, due to labor-induced losses of key intermediate, concentrated components of the manufactured food supply chain and to labor-induced losses of transportation services. Either loss could be sufficient to cause food shortages in local supermarkets, which could start self-reinforcing viral waves of fear, hoarding, sharp increases in demand, household stress, and then the herein-described emotion-based economic behaviors.

## 4.0 Summary and Conclusions

This paper describes a new cognitive-economic model of household purchasing behavior developed to overcome a number of known limitations of traditional economic modeling. The model behind this cognition-based household purchasing is a semantic memory network, composed of external environmental cues, internal contexts that are activated by the relative values of these cues, and action intentions that result from the activation levels of the internal contexts. The model captures how a high state of fear (e.g., during an extreme event) causes a household to make decisions differently than in a low state of fear (e.g., during normal conditions). If the set of particular contexts results in a state of low fear, then the semantic network uses a particular set of activation levels; if the set of contexts result in a state of high fear, it uses another. Finally, the model accounts for the known differences in behaviors of households with high socio-economic status versus low socio-economic status, by using different sets of network activation edges and weights.

To see how this cognitive-economic household model impacts the distribution of food in an economy during an extreme event, simulations were conducted of the impacts of a pandemic influenza on the psychological behavior of households and ultimately their food purchasing behavior. The most common phenomenon observed in the simulations is that at the onset of the disease in a household, this household will experience high levels of fear, invoking the high-fear semantic network between contexts and action intentions, thereby stockpiling food and then self-isolating. Once other households observe this stockpiling and self-isolation, they also attain high levels of fear, resulting in stockpiling as well. If high levels of stockpiling occur, then other, unaffected households will observe large reductions in stock at their supermarket, invoking them into high fear levels as well. Said differently, the spread of fear and economic hoarding behavior is much faster than the spread of the disease itself.

The primary finding of these simulations is that many of the possible sets of realistic pandemic conditions can cause severe shortages in local food supply above and beyond those caused by the disease itself. While traditional economic models would suggest that the rate of food purchases would remain constant through a pandemic, the simulations demonstrate that food stockpiling can occur rapidly and non-linearly, and be caused by any number of events (illness, widespread panic, or empty food shelves).

The most important conclusion to be drawn from this work is that the inherent, multiplicative effects of the infectious network of pandemic spread, the neighborhood networks of information about illness, stockpiling, and supermarket shortages, and the network effects of food supply and demand exacerbate the actions within a single household. Much like the irrational fears that cause bank runs on monetary deposits, households' irrational cognition regarding the pandemic and its commodity effects cause them to create supermarket runs on food. In addition to rationing food, government policies should consider potential means of "rationing" fear; that is, ensuring there is enough warranted fear for households to have sufficient motivation and food to self-isolate, thereby mitigating the disease, but not so much as to cause widespread fear that in and of itself causes widespread food shortages.

# References

1. Canadian Supply Chain Food Safety Coalition (2007) Pandemic Influenza Emergency Simulation Project for the Agri-Food Sector. The Zeta Group, Ottawa, Ontario.
2. City of Albuquerque (2008) GIS Data from the City of Albuquerque. Albuquerque, NM.
3. The D&B Corporation (2007) Short Hills, NJ.
4. Ehlen M.A., Downes P.S., Scholand, A.J. (October 2007) Economic Impacts of Pandemic Influenza on the U.S. Manufactured Foods Industry. DHS, NISAC, Albuquerque, NM.
5. Ehlen, M.A., Downes, P.S., Scholand, A.J. (March 2007) A Post-Katrina Comparative Economic Analysis of the Chemicals, Food, and Goodyear Value Chains (OUO). DHS, Albuquerque, NM.
6. Eidson, E.D., Ehlen, M.A. (2005) NISAC Agent-Based Laboratory for Economics (N-ABLE), SAND2005-0263. Sandia National Laboratories, Albuquerque, NM.
7. The Food Marketing Institute (2006) Avian Influenza & Pandemic Preparedness. Arlington, VA.
8. Florida Department of Agriculture and Consumer Services (2008). Pandemic Influenza Agriculture Planning Toolkit. Tallahassee, FL.
9. Kay V., Smith B., Hoyt T. et al (2007). The Role of Emotion in Decision Making in an Avian Flu. Submitted to American Journal of Public Health.
10. Smith B.W., Kay V. S., Hoyt T. et al (2007) The Effects of Personality and Emotion on Behavior in a Potential Avian Flu Epidemic. 11th International Conference on Social Stress Research.
11. The White House (2005) National Strategy for a Pandemic Influenza. Washington, D.C.

# Collaborating with Multiple Distributed Perspectives and Memories

Anthony J. Ford[†] and Alice M. Mulvehill[*]

[†] Anthony.Ford@rl.af.mil, AFRL, Rome NY
[*] amm@bbn.com, BBN Technologies, Cambridge MA

**Abstract** Experience-based reasoning is a powerful tool for decision-support systems which attempt to leverage human memories as examples to solve problems. However, there is also a requirement to leverage recorded experiences that are created by a set of diverse users from a variety of work positions, with varied goals, cultural backgrounds, and perspectives across time. New computing methods need to be specified and defined in order to resolve some of the problems introduced by these requirements. The research discussed in this paper focuses on the ongoing issue of how to reconcile the different personal and cultural perspectives of agents in order to make group decisions. We discuss how individual agent history and culture can influence experience-based reasoning. We compare our approaches to traditional experience-based reasoning techniques in order to highlight important issues and future directions in social decision making. Results from an experiment are presented that describe how shared problem solving presents issues of consensus and buy-in for various stakeholders and how a constraint based coherence mechanism can be used to ensure coherence in joint action among a group of agents and improve performance.

## 1 Introduction

People *learn* to communicate about individual and shared experiences. Shared psychophysical capabilities and formal education form cultural models that insure that our perceptions are related in a manner that enables people to share information. At the same time, there are factors at work that lead to different perceptions of the same events. Experience-based reasoning is a powerful tool for decision-support systems which attempt to leverage human memories as examples to solve problems [7, 13]. The fine details of how culture-based biases affect retrieval, classification and problem solving are areas of active research investigation [11, 8].

Internet technology and availability is changing the social culture of information access and exchange. With this technology, there are benefits e.g., information is available on demand and from any location where there is a portal; and negative aspects e.g., issues of information integrity can make it difficult for the end user to have

H. Liu et al. (eds.), *Social Computing and Behavioral Modeling*,
DOI: 10.1007/978-1-4419-0056-2_12, © Springer Science + Business Media, LLC 2009

trust in what is available [2]. For military systems, the Network Centric Warfare concept [18] is beginning to transform the information networks and services developed to support users. As new network centric applications become available, users who leverage the network centric concept are themselves evolving a new culture of information access and exchange. Distributed computing is becoming a norm.

The work discussed in this paper highlights some of the issues surrounding cross-cultural experience exchange and presents current research into some of these issues. Specifically, we discuss how the goals, culture, role, and history of an agent can influence how it exchanges and uses past experiences to solve problems in a group setting. We present the task of multi-agent planning as an example for our techniques, but we intend our approaches to these problems to be applicable to a variety of problems faced by both human and multi-agent systems.

## 2 Distributed Episodic Memory

Distributed Episodic Exploratory Planning (DEEP) is an AFRL program that is developing a mixed-initiative decision-support environment where commanders can readily access and leverage *historical* data from distributed sources for use in decision making. Episodic reasoning paradigms and specific case based reasoning (CBR) technology are being considered because of the requirements for analogical reasoning and learning from past experience [4]. Researchers in the field of CBR have suggested two dimensions along which distributed approaches to CBR systems can be defined: how knowledge is processed in the system and how knowledge is organized in the system [14]. The processing may be accomplished by a single agent or by multiple agents and the knowledge may be organized within a single or multiple case base(s).

In humans, previous experiences are often referred to as episodic memories. Annotation and storage of the memories tend to be highly individualized and self oriented. The content for a typical automated episodic reasoning system or case base is the recorded experiences of a particular operator or the collected experiences from a set of operators in a given problem solving context.

In coordination with DEEP, the DEAR (Distributed Episodic Analogical Reasoning) project is conducting research that is focused specifically on issues associated with the use of episodes and analogies, by both humans and software agents, in distributed group interaction and collaboration. To date, our research has been influenced by: the work of Gentner [5] on mental models and analogical reasoning; the work on Shank [15], Leake [10] and Kolodner [9] on case base reasoning; the work of Gasser [6] on distributed artificial intelligence; the work of Endsley [3] on situation awareness; and the work of Mantovani [12] on actor-environment interaction.

In the next sections, we highlight issues surrounding cross-cultural experience exchange and present research findings that indicate how the role, goals, and cultural history of an entity influences how it performs experience-based reasoning in a group setting. Since the disparity of experiences in a group setting can impact decision-making, our research advocates a coherence maintenance mechanism to improve performance. In this paper, we describe a constraint satisfaction approach and present an implementation and some early experimental results that describe the benefits of the approach. We conclude with observations and directions for future work.

## 3 Memory and Cultural Models

Should cultural models influence what features of an experience are important during recall? For example, say a commander in a command center had previously served as an intelligence officer. Our research indicates that the experience and memories accumulated in the previous role as an intelligence officer will influence how the commander communicates with the individual holding that intelligence role in his current situation. Perhaps the commander will be able to better understand the presentation of the intelligence officer, better appreciate the differences in the source and integrity of the information presented, or perhaps he will have certain expectations of what the intelligence officer in the current position should be providing.

What does it mean to have "distributed memory"? Is this some type of "collective memory"? If so, would the memory of one entity about a certain event be the same as the memory of that same event by another entity? The pattern recognition literature would suggest that this is unlikely. Instead the memory would be a function of the sensor and the feature extraction of the receiving entity. The findings in the psychological literature are similar—memory is described as a subjective experience. That is, each person participating in an event will attend to and remember particular aspects of the event dependent on the role they are playing in the action. The driver of a car and a pedestrian will each have a different view of an accident; an operations officer and an intelligence officer will each take away a different view of a military engagement. We each build an episodic memory, but from our perspective of what happened. But because we accumulate experience and learn (both autonomously and by being told), we probably have larger models that enable us to know how to refer to the "same story".

So, how would this distributed memory be stored and accessed? Would it be stored as one problem with multiple views or perspectives? For example, in the case of the accident would there be one episode with two views, or would there be two separate episodes with features such as time and space that could be used to relate them to each other? Does the method selected for storage affect retrieval? For example, if the accident is stored as two separate episodes, one for each participant, then during retrieval from the distributed cases, how would the receiver determine that the retrieved episodes were: a) two memories for the same event, or b) one accurate and one inaccurate memory? Without the overarching information of the more general problem (the accident), this becomes a hard issue to resolve. If there were a third set of observations, such as those from a police officer who recorded the observations from all of the actors at the accident scene, would this help resolve some of the ambiguities by providing a larger context? Reuse is highly influenced by the current goal of the actor in the current problem solving situation. How the multiple perspectives of a single episode can be used effectively is a difficult research topic. Think of the way a police officer has to listen to every side of the story before deciding 'what happened'. What tools and techniques (protocols and/or policies) does the officer have available to him or her for examining different versions of events and for relating events from disparate sources in a way that will ensure a coherent story?

These are examples that highlight the need for a mechanism that can reconcile multiple perspectives while leveraging group diversity and establishing a shared understanding by unique participants and/or observers. In the next section, we will show how an agent's cultural model influences how it retrieves and shares experiences which are relevant to a current problem. We will present an approach that describes experience in shared problem solving from the perspective of participants, each with a role, past history associated with the role, and information about the cultural model to which the agent prescribes. An agent's cultural model will lead to different interpretations of features in the environment, influencing how it utilizes its past history.

# 4 Maintaining Coherence

Like the police officer in the previous example, our goal is to identify methods that merge multiple perspectives about a given problem into one coherent solution. We hypothesize that in order for a rational solution to evolve in a distributed setting, some mechanism needs to be in place to maintain coherence. Coherence, in this context, refers to a source of justification for information which relies on the relationships between individual pieces of information. Each piece of information is only justified in the grand scheme of the overall system [1].

Coherence has been described computationally as a constraint satisfaction problem by Thagard and Verbeurgt [16] and as a method of making decisions by Thagard and Millgram [17]. These models of coherence involve using constraints to describe the coherence and incoherence among actions and goals. Coherence is established by selecting the most coherent set of actions and goals to undertake. This approach has been extended by Ford and Lawton [4] to include falsification as a perturbation to coherence to increase the likelihood that coherent actions and goals also correspond to current evidence.

To test our hypothesis, we have implemented a coherence monitoring agent that is responsible for discovering how the proposed actions and goals of a set of agents are related to each other in a system of coherence. In the implementation, multiple agents contribute actions and goals from experience when faced with a problem, based on each agent's history and cultural model as described above. The agents use coherence and falsification to select a collective plan by selecting the most coherent (and least falsified) set of actions and goals to undertake. In order to enhance this approach we take cultural models into consideration when forming coherent and incoherent relationships between actions or goals.

Each agent assumes a role within an overall culture. This role describes the important features for recall and planning, and allows each agent to utilize their personal history in a culturally-informed way. The coherence monitoring agent discovers consistencies between actions and goals based on the case structure of each individual agent. The coherence monitoring agent also discovers inconsistencies by comparing the resources used and policies undertaken to find overlap. The specific details of our cultural models and test scenario are outlined in the following section.

## 5 Experimental Results

In order to test our hypotheses regarding culturally-informed distributed memory, we implemented a multi-agent system which models different approaches to the problem of stabilizing a fictitious nation-state. Nation-state stabilization is a complicated, multi-variant issue requiring multiple (possibly competing) goals to be undertaken at once. We hypothesize that a diverse, multi-cultural approach will outperform a monolithic case-based reasoner in such a domain. For this problem, we required that our agents develop policies to undertake which determine specific investments in a fictional nation's security and economy. Which investments are most advantageous is a matter of perspective, and so we implemented various cultural models to allow for different perspectives on advantageous policies to undertake and their indicators of success in the environment.

The cultural models we implemented each contained different roles which parameterize the behavior of adherent agents. These roles change how each agent recalls, adapts, and stores experiences in its personal history. To this end, each role includes three sets of variables relevant to the test scenario. The first set is the key situation factors, which constrain how the agent searches its personal history for past experiences to utilize. The second set is the policy variables, which constrain what variables each agent is expected to change in the environment. The third set is the expected outcome set, which constrains how each agent assesses success in the environment after executing their policy.

For our testing, we implemented two cultural models, each with two roles. The first model we implemented was a military advisor model. This model included two roles: traditional military trainers and counterinsurgency advisors. These roles are both concerned with improving the host-nation's military readiness, but use different policies and indicators to achieve that end. The traditional military trainer focuses on indigenous military forces and their casualty rates, while the counterinsurgency advisor focuses more on border patrol and infrastructure security.

The second cultural model we implemented was an economic policy advisor culture. Within that culture, we described two roles: rural development advisors and urban development advisors. These roles are both concerned with enhancing the economic health of a host-nation, but differ in the approaches and indicators they employ. The rural development advisor focuses on agricultural sector health and national economic health, while the urban development advisor focuses on industrial and service sector health and the economic health of urban areas.

Much like the traffic accident example above, the four advisors have different views of events. The military development advisors may emphasize increasing the size of military forces to address unemployment, while the economic advisors may emphasize private industry stimulus as a way to address the same issue. Moreover, the traditional military advisor may use force size to indicate success while the economic advisors observe GDP (gross domestic product) growth rate to indicate success. Like the police officer in the previous example, the coherence monitoring agent must determine the most plausible policy to undertake even though the different perspectives may not agree.

To gather results on the performance of our different agents, we utilized an in-house simulation environment which models various facets of a fictitious nation-state. These facets are represented by numeric variables which interact over time. In order to test the policies our agents individually and collectively developed, we set the model into an initial state of instability, and employed our multi-agent system to develop a set of policy variables to undertake. In our test nation model, we had three goals: to reduce the murder rate, to reduce the unemployment rate, and to increase the GDP growth rate.

Each agent adhered to a cultural model and a role within it, so a total of four agents represented the different perspectives explained above. In order to develop a shared policy to undertake, each agent provided an individual policy based on their personal histories. A coherence agent was responsible for enforcing coherence in the final result, combing the four unique policies into one collective approach.

To compare the collective and individual policies, we established the baseline performance of the model and measured the impact each policy had on those three goals listed above (lower unemployment, lower murder rates, and higher GDP growth) we determined the performance of the policies by comparing their impact on those three variables in the model. Table 1 presents our results:

**Table 1.** Final values for the test model.

| Criterion | Best Economic Dev. Plan | Best Military Dev. Plan | Collective Plan |
|---|---|---|---|
| Urban GDP Growth | 2.47% | 2.43% | 2.36% |
| Rural GDP Growth | 2.47% | 2.48% | 2.47% |
| Unemployment | 25.65% | 26.10% | 26.23% |
| Murder Rate per 100k | 38.53 | 40.21 | 40.84 |

Unfortunately, our results show the combination of policies was disadvantageous. Although the differences in performance were not significant, we should still consider why the coherent plan took on poor characteristics from the individual plans of which it was comprised. A possible explanation is the failure of the individual experiences to capture more sophisticated causal linkages between input and output features. The examination of coherence in the *beliefs* in causality may lead to better performance, since the agents would then have to reconcile their approaches to problems based on their beliefs about the impact their approaches would have. Simply recording the input and output factors and assuming a causal relationship may have left important inconsistencies unnoticed. These results will allow us to further explore and refine our implementation of culturally-informed deliberative coherence.

# 6 Conclusion

The work discussed in this paper has highlighted some of the issues surrounding cross-cultural experience exchange and presented current research into some of these issues. To date, we have developed an approach that describes experience in shared problem solving from the aspect of participants, each with a role, past history associated with the role, and information about the cultural model that described the policies and procedures for the role within the specific problem solving domain. The use of shared experiences by multiple distributed humans or automated services on computational systems is still being investigated. We recognize that different cultures might have different assumptions about the structure, content, causality, and usefulness of experience in problem solving. In our future work, we will enhance the coherence monitoring agent to include data on how different cultures assess and analyze their own decision making in order to better reconcile the actions and goals which make up the solution to a shared problem.

Our preliminary attempts at reconciling disparate experiences must be re-examined and amended based on the issues highlighted earlier with the differences in perspective, culture, and history between agents. Our future work will allow for multiple agents to *learn* how to establish coherence based on their individual and cultural perspectives.

# References

1. Bonjour, Lawrence. (1976). The Coherence Theory of Empirical Knowledge. Philosophical Studies, 30 (5), pp. 281-312.
2. Brey, P. (2005). Evaluating the social and cultural implications of the Internet, *ACM SIGCAS Computers and Society, 35,* 3.
3. Endsley, M. R. & Garland, D. J. (Eds.) (2000). *Situation awareness analysis and measurement.* Mahwah, NJ: Lawrence Erlbaum Associates.
4. Ford, A. J., & Lawton, J. H. (2008). Synthesizing Disparate Experiences in Episodic Planning. Proceedings of the 13th ICCRTS: C2 for Complex Endeavors.
5. Gentner, D. & Stevens, A. L. (1983). In D. Gentner & A. L. Stevens (Eds.), *Mental models,* (pp. 1-6). Hillsdale, NJ: Lawrence Erlbaum.
6. Gasser, Les, (1991) Social Conceptions of Knowledge and Action: DAI Foundations and Open Systems Semantics, Artificial Intelligence, Vol. 47, pp. 107 – 138
7. Hammond, Kristian J, Mark J Fasciano, Daniel D Fu, & Timothy Converse (1996). Actualized Intelligence: Case-based Agency in Practice. Applied Cognitive Psychology, 10. S73-S83.
8. Javidan, M., House, R. J., & Dorfman P. W. (2004). Summary of The GLOBE Study. In R. J. House, P. J. Hanges, M. Javidan, P. W.. Dorfman, & V. Gupta (Eds.), *Culture, Leadership and Organizations: Globe Study of 62 Societies* (pp. 9-27). Thousand Oaks, CA: Sage Publications.
9. Kolodner, J. (1993). *Case-Based Reasoning.* San Mateo, CA: Morgan Kaufmann.
10. Leake, D. B. (Ed.) (1996). *Case Based Reasoning: Experiences, Lessons and Future Directions.* Cambridge, MA: MIT Press.
11. Lonner, W. J. Dinnel, D. L. Hayes, S. A. & Sattler, D. N. (Eds.). (2007). *Online readings in psychology and culture.* http://www.ac.wwu.edu/~culture/readings.htm.
12. Mantovani, Giusepe, Social Context in HCI: A New Framework for Mental Models, Cooperation, and Communication, (1996) Cognitive Science, Vo. 20, No. 2, pp. 237 – 270
13. Mulvehill, A., (1996) "Building, Remembering, and Revising Force Deployment Plans", *Advanced Planning Technology - Technological Achievements of the ARPA, Rome Laboratory Planning Initiative,* AAAI Press.
14. Plaza, E. & McGinty, L. (2005). Distributed case-based reasoning. *The Knowledge Engineering Review, 20,* 261-265.
15. Schank, R. C. (1982). *Dynamic memory: A theory of reminding and learning in computers and people.* New York, NY: Cambridge University Press.
16. Thagard, P and K Verbeurgt (1998). Coherence as Constraint Satisfaction. Cognitive Science, 22(1). 1-24
17. Thagard, P. and Millgram, E. (1995) Inference to the best plan: A coherence theory of decision. In A. Ram & D. B. Leake (Eds.), Goal-driven learning.
18. Wilson, C. (2004). *Network Centric Warfare: Background and Oversight Issues for Congress,* Order Code RL32411, Congressional Research Service, the Library of Congress.

# When is social computation better than the sum of its parts?

Vadas Gintautas[1], Aric Hagberg[2], and Luís M. A. Bettencourt[3]

Center for Nonlinear Studies, and Applied Mathematics and Plasma Physics,
Theoretical Division, Los Alamos National Laboratory, Los Alamos NM 87545
[1]vadasg@lanl.gov, [2]hagberg@lanl.gov, [3]lmbett@lanl.gov

**Abstract** Social computation, whether in the form of searches performed by swarms of agents or collective predictions of markets, often supplies remarkably good solutions to complex problems. In many examples, individuals trying to solve a problem locally can aggregate their information and work together to arrive at a superior global solution. This suggests that there may be general principles of information aggregation and coordination that can transcend particular applications. Here we show that the general structure of this problem can be cast in terms of information theory and derive mathematical conditions that lead to optimal multi-agent searches. Specifically, we illustrate the problem in terms of local search algorithms for autonomous agents looking for the spatial location of a stochastic source. We explore the types of search problems, defined in terms of the statistical properties of the source and the nature of measurements at each agent, for which coordination among multiple searchers yields an advantage beyond that gained by having the same number of independent searchers. We show that effective coordination corresponds to synergy and that ineffective coordination corresponds to independence as defined using information theory. We classify explicit types of sources in terms of their potential for synergy. We show that sources that emit uncorrelated signals provide no opportunity for synergetic coordination while sources that emit signals that are correlated in some way, do allow for strong synergy between searchers. These general considerations are crucial for designing optimal algorithms for particular search problems in real world settings.

## 1 Introduction

The ability of agents to share information and to coordinate actions and decisions can provide significant practical advantages in real-world searches. Whether the target is a person trapped by an avalanche or a hidden cache of nuclear material, being able to deploy multiple autonomous searchers can be more advantageous and safer than sending human operators. For example, small autonomous, possibly expendable robots could be utilized in harsh winter climates or on the battlefield.

H. Liu et al. (eds.), *Social Computing and Behavioral Modeling*,
DOI: 10.1007/978-1-4419-0056-2_13, © Springer Science + Business Media, LLC 2009

In some problems, e.g. locating a cellular telephone via the signal strength at several towers, there is often a simple geometrical search strategy, such as triangulation, which works effectively. However, in search problems where the signal is stochastic or no geometrical solution is known, e.g. searching for a weak scent source in a turbulent medium, new methods need to be developed. This is especially true when designing autonomous and self-repairing algorithms for robotic agents [1]. Information theoretical methods provide a promising approach to develop objective functions and search algorithms to fill this gap. In a recent paper, Vergassola et al. demonstrated that infotaxis, which is motion based on expected information gain, can be a more effective search strategy when the source signal is weak than conventional methods such as moving along the gradient of a chemical concentration [2]. The infotaxis algorithm combines the two competing goals of exploration of possible search moves and exploitation of received signals to guide the searcher in the direction with the highest probability of finding the source [3].

To improve the efficiency of search by using more than one searcher requires determining under what circumstances a collective (parallel) search is better (faster) than the independent *combination* of the individual searches. Much heuristic work, anecdotally inspired by strategies in social insects and flocking birds [4, 5], has suggested that collective action should be advantageous in searches in real world complex problems, such as foraging, spatial mapping, and navigation. However, all approaches to date rely on simple heuristics that fail to make explicit the general informational advantages of such strategies.

The simplest extension of infotaxis to collective searches is to have multiple *independent* (uncoordinated) searchers that share information; this corresponds in general to a linear increase in performance with the number of searchers. However, given some general knowledge about the structure of the search, substantial increases in the search performance of a collective of agents can be achieved, often leading to exponential reduction in the search effort, in terms of time, energy or number of steps [6, 7, 8]. In this work we explore how the concept of information synergy can be leveraged to improve infotaxis of multiple coordinated searchers. Synergy corresponds to the general situation when measuring two or more variables *together* with respect to another (the target's signal) results in a greater information gain than the sum of that from each variable *separately* [9, 10]. We identify the types of spatial search problems for which coordination among multiple searchers is effective (synergetic), as well as when it is ineffective, and corresponds to independence. We find that classes of statistical sources, such as those that emit uncorrelated signals (e.g. Poisson processes) provide no opportunity for synergetic coordination. On the other hand, sources that emit particles with spatial, temporal, or categorical correlations, do allow for strong synergy between searchers that can be exploited via coordinated motion. These considerations divide collective search problems into different general classes and are crucial for designing effective algorithms for particular applications.

## 2 Information theory approach to stochastic search

Effective and robust search methods for the location of stochastic sources must balance the competing strategies of exploration and exploitation [3]. On the one hand, searchers must exploit measured cues to guide their optimal next move. On the other hand, because this information is statistical, more measurements need to typically be made that are guided by different search scenarios. Information theory approaches to search achieve this balance by utilizing movement strategies that increase the expected information gain, which in turn is a functional of the many possible source locations. In this section we define the necessary formalism and use it to set up the general structure of the stochastic search problem.

### 2.1 Synergy and Redundancy

First we define the concepts of information, synergy and redundancy explicitly. Consider the stochastic variables $X_i, i = 1 \ldots n$. Each variable $X_i$ can take on specific states, denoted by the corresponding lowercase letter, that is $X$ can take on a set of states $\{x\}$. For a single variable $X$ the Shannon entropy (henceforth "entropy") is $S(X) = -\sum_x P(x) \log_2 P(x)$, where $P(x)$ is the probability that the variable $X$ take on the value $x$ [11]. The entropy is a measure of uncertainty about the state of $X$, therefore entropy can only decrease or remain unchanged as more variables are measured. The conditional entropy of a variable $X_1$ given a second variable $X_2$ is $S(X_1|X_2) = -\sum_{x_1,x_2} P(x_1,x_2) \log_2 (P(x_1,x_2)/P(x_2)) \leq S(X_1)$. The mutual information between two variables, which plays an important role in search strategy, is defined as the change in entropy when a variable is measured $I(X_1,X_2) = S(X_1) - S(X_1|X_2) \geq 0$. These definitions can be directly extended to multiple variables. For 3 variables, we make the following definition [12]: $R(X_1,X_2,X_3) \equiv I(X_1,X_2) - I(\{X_1,X_2\}|X_3)$. This quantity measures the degree of "overlap" in the information contained in variables $X_1$ and $X_2$ with respect to $X_3$. If $R(X_1,X_2,X_3) > 0$, there is overlap and $X_1$ and $X_2$ are said to be redundant with respect to $X_3$. If $R(X_1,X_2,X_3) < 0$, more information is available when these variables are considered together than when considered separately. In this case $X_1$ and $X_2$ are said to be synergetic with respect to $X_3$. If $R(X_1,X_2,X_3) = 0$, $X_1$ and $X_2$ are independent [9, 10].

### 2.2 Two-dimensional spatial search

We now formulate the two-dimensional stochastic search problem. We consider, for simplicity, the case of two searchers seeking to find a stochastic source located in a finite two-dimensional plane. This is a generalization of the single searcher formalism presented in Ref. [2]. At any time step, the searchers have positions $\{r_i\}, i = 1,2$ and observe some number of particles $\{h_i\}$ from the source. The searchers do not get information about the trajectories or speed of the particles; they only get information if a particle was observed or not. Therefore simple geometrical methods such as triangulation are not possible. Let the variable $R_0$ correspond to all the possible locations of the source $r_0$. The searchers compute and share a probability distribution $P^{(t)}(r_0)$ for the source at each time index $t$. Initially the probability for the source is assumed to be to be uniform. After each measurement $\{h_i, r_i\}$, the searchers update their estimated probability distribution of source positions via Bayesian inference. First the conditional probability $P^{(t+1)}(r_0|\{h_i, r_i\}) \equiv P^{(t)}(r_0)P(\{h_i, r_i\}|r_0)/A$, is calculated, where $A$ is a normalization over all possible source locations as required by Bayesian inference. This is then assimilated via Bayesian update so that $P^{(t+1)}(r_0) \equiv P^{(t+1)}(r_0|\{h_i, r_i\})$.

If the searchers do not find the source at their present locations they choose the next local move using an infotaxis step to maximize the expected information gain. To describe the infotaxis step we first need some definitions. The entropy of the distribution $P^{(t)}(r_0)$ at time $t$ is defined as $S^{(t)}(R_0) \equiv -\sum_{r_0} P^{(t)}(r_0) \log_2 P^{(t)}(r_0)$. In terms of a specific measurement $\{h_i, r_i\}$ the entropy is (*before* the Bayesian update) $S^{(t)}_{\{h_i, r_i\}}(R_0) \equiv -\sum_{r_0} P^{(t)}(r_0|\{h_i, r_i\}) \log_2 P^{(t)}(r_0|\{h_i, r_i\})$. We define the difference between the entropy at time $t$ and the entropy at time $t+1$ after a measurement $\{h_i, r_i\}$ to be $\Delta S^{(t+1)}_{\{h_i, r_i\}} \equiv S^{(t+1)}_{\{h_i, r_i\}}(R_0) - S^{(t)}(R_0)$.

Initially the entropy is at its maximum for a uniform prior: $S^{(0)}(R_0) = \log_2 N_s$, where $N_s$ is the number of possible locations for the source in a discrete space. For each possible joint move $\{r_i\}$, the change in expected entropy $\overline{\Delta S}$ is computed and the move with the minimum (most negative) $\overline{\Delta S}$ is executed. The expected entropy is computed by considering the reduction in entropy for all of the possible joint moves

$$\overline{\Delta S} = - \left[ \sum_i P^{(t)}(R_0 = r_i) \right] S^{(t)}(R_0)$$
$$+ \left[ 1 - \sum_i P^{(t)}(R_0 = r_i) \right] \Delta S^{(t+1)}_{\{h_i, r_i\}} \sum_{h_1, h_2} \left[ \sum_{r_0} P^{(t)}(r_0) P^{(t+1)}(\{h_i, r_i\}|r_0) \right]. \quad (1)$$

The first term in Eq. (1) corresponds to one of the searchers finding the source in the next time step (the final entropy will be $S = 0$ so $\overline{\Delta S} = -S$). The second term considers the reduction in entropy for all possible measurements at the proposed location, weighted by the probability of each of those measurements. The probability of the searchers obtaining the measurement $\{h_i\}$ at the location $\{r_i\}$ is given by the trace of the probability $P^{(t+1)}(\{h_i, r_i\}|r_0)$ over all possible source locations.

## 3 Correlated stochastic source and synergy of searchers

The expected entropy reduction $\overline{\Delta S}$ is calculated for *joint* moves of the searchers, that is, all possible combinations of individual moves. Compared with multiple independent searchers this calculation incurs some extra computational cost. Thus, when designing a search algorithm, it is important to know whether an advantage (synergy) can be gained by considering joint moves instead of individual moves. Since the search is based on optimizing the maximum information gain we need to explore if joint moves are synergetic or redundant. In this section we will show how correlations in the source affect the synergy and redundancy of the search.

In the following we will assume there are no radial correlations between particles emitted from the source and that the probability of detecting particles decays with distance to the source. For each particle emitted from the source, the searcher $i$ has an associated actual probability $\pi_i(r_0)$ of catching the particle. The probability $\pi_i(r_0)$ is defined in terms of a possible source location $r_0$ and the location $r_i$ of searcher $i$: $\pi_i(r_0) = B\exp(-|\mathbf{r}_i - \mathbf{r}_0|^2)$, where $\{r_i\}$ is the set of all the searcher positions and $B$ is a normalization constant. Note that this is just the radial component of the probability; if there are angular correlations these are treated separately. We may now write $R$, as a function of the variables $R_0$, $H_1$, and $H_2$, in terms of the conditional probabilities:

$$R(H_1, H_2, R_0) = \sum_{h_1, h_2, r_0} P(r_0, h_1, h_2) \log_2 \frac{P(h_1|r_0)P(h_2|r_0)P(h_2|h_1)}{P(h_2)P(h_1, h_2|r_0)}. \qquad (2)$$

It is sufficient for $R \neq 0$ that the argument of the logarithm differs from 1. This can be achieved even if measurements are conditionally independent (redundancy), mutually independent (synergy), or when neither of these conditions apply.

### 3.1 Uncorrelated signals: Poisson source

First, consider a source which emits particles according to a Poisson process with known mean $\lambda_0$ so emitted particles are completely uncorrelated spatially and temporally. If searcher 1 is able to get a particle that has already been detected by searcher 2, it is clear that the searchers are completely independent and there is no chance of synergy. It may appear at first that implementing a simple exclusion where two searchers cannot get the same particle would be enough to foster cooperation between searchers. We will instead show that it is the Poisson nature of the source that makes synergy impossible, even under mutual exclusion of the measurements.

The probability of the measurement $\{h_i\}$ is given by

$$P(\{h_i, r_i\}|r_0) = \sum_{h_s = \sum_i h_i}^{\infty} P_0(h_s, \lambda_0) M(\{\pi_i(r_0)\}, \{h_i\}, h_s). \qquad (3)$$

The sum is over all possible values of $h_s$, weighted by the Poisson probability mass function with the known mean $\lambda_0$. $M$ is the probability mass function of the multinomial distribution for that measurement; it handles the combinatorial degeneracy and the exclusion. It is not difficult to show by summing over $h_s$ that $P(\{h_i, r_i\}|r_0)$ can be written as a product of Poisson distributions with effective means $\lambda_0 \pi_i$,

$$P(\{h_i, r_i\}|r_0) = \frac{\lambda_0^{\Sigma_i h_i} e^{-\lambda_0 \Sigma_i \pi_i} \prod_i \pi_i^{h_i}}{\prod_i (h_i!)} = \prod_i P_0(h_i, \lambda_0 \pi_i). \tag{4}$$

At this point we consider whether a search like this can be synergetic for the 2 searcher case. Eq. (4) shows that the two measurements are conditionally independent and therefore $P(h_1, h_2|r_0) = P(h_1|r_0)P(h_2|r_0)$. It follows from Eq. (2) that $R(H_1, H_2, R_0) = I(H_1, H_2) \geq 0$. Therefore the searchers are either redundant (if the measurements interfere with each other) or independent with respect to the source. Synergy is impossible so that searchers gain no advantage by considering joint moves. The only advantage of coordination comes possibly from avoiding positions that lead to a decrease in performance of the collective due to competition for the same signal.

## 3.2 Correlated signals: angular biases

We now consider a source that emits particles that are spatially correlated. We assume for simplicity that at each time step the source emits 2 particles. The first particle is emitted at a random angle $\theta_{h_1}$ chosen uniformly from $[0, 2\pi)$. The second particle is emitted at an angle $\theta_{h_2}$ with probability

$$P(\theta_{h_2}|\theta_{h_1}) = D \exp\left[-(|\theta_{h_1} - \theta_{h_2}| - \pi)^2/\sigma^2\right] \equiv f, \tag{5}$$

where $D$ is a normalization factor. The searchers are assumed to know the variance $\sigma$ for simplicity; this is a reasonable assumption if the searchers have any information about the nature of the target (just as for the Poisson source they had statistical knowledge of the parameter $\lambda_0$). The calculation of the conditional probability $P(\{h_i, r_i\}|r_0)$ requires some care. Specifically, this quantity is the probability of the measurement $\{h_i\}$, assuming a certain source position. Since there are 2 particles emitted at each time step, there are 4 possible cases, each with a different probability, as shown in Table 1. Here $\theta_{h_1}$ and $\theta_{h_2}$ are calculated from $r_1$ and $r_2$, respectively: $\theta_{h_i} \equiv \arctan \frac{r_{0,y} - r_{i,y}}{r_{0,x} - r_{i,x}}$. Note that the $\pi_i$ are functions of $r_0$. The coefficient $D$ is chosen such that $\frac{1}{N} \Sigma_{h_2} \Sigma_{r_2} P(\{h_1, h_2, r_1, r_2\}|r_0) = P(\{h_1, r_1\}|r_0)$, corresponding to the normalization condition $\frac{1}{N} \Sigma_{r_2} f = 1$.

Figure 1 shows the value of $R(H_1, H_2, R_0)$ and the values of the mutual informations $I(H_1, H_2)$ and $I(H_1, H_2|R_0)$ for each possible position $r_2$ of searcher 2. We assume a nonuniform, peaked probability distribution for the source [Figure 1(b)] and that the position of searcher 1 is fixed. In this setup we see that $R <= 0$ for ev-

| $\{h_1,h_2\}$ | $\{1,1\}$ | $\{1,0\}$ | $\{0,1\}$ | $\{0,0\}$ |
|---|---|---|---|---|
| $P(\{h_1,r_1\}\vert r_0)$ | $\pi_1$ | $\pi_1$ | $1-\pi_1$ | $1-\pi_1$ |
| $P(\{h_2,r_2\}\vert r_0)$ | $\pi_2$ | $\pi_2$ | $1-\pi_2$ | $1-\pi_2$ |
| $P(\{h_1,h_2,r_1,r_2\}\vert r_0)$ | $\pi_1\pi_2 f^2$ | $\pi_1 f(1-\pi_2 f)$ | $\pi_2 f(1-\pi_1 f)$ | $(1-\pi_1 f)(1-\pi_2 f)$ |

**Table 1** Probability calculation for all possible states in the correlated source search. Here $\pi_i(r_0) = B\exp\left(-\vert \mathbf{r}_i - \mathbf{r_0}\vert^2\right)$ is written as $\pi_i$ to save space.

ery possible position of searcher 2 indicating that only synergy is possible. This is a consequence of the angular spatial correlation between the particles emitted by the source. The synergy is highest near the source location, where the source probability is strongly peaked, and falls off rapidly away from the source location. Furthermore there is little to no synergy near searcher 1 since in that region it is very unlikely that both searchers would simultaneously observe a particle. The area of greatest synergy corresponds to the most probable source locations for both searchers to simultaneously observe a particle. $P(r_0)$ is very flat at the boundaries; thus $R_0$ contributes little to $I(H_1, H_2 \vert R_0)$ in the lower left corner and $R$ is small.

## 4 Conclusion

In the real world, communication between agents, as well as centralized or decentralized real-time computation can be difficult or expensive. Therefore it is important to consider the classes of search problems for which coordination between searchers can achieve quantitative advantages over independent agents. In this work we studied search algorithms for autonomous agents looking for the spatial location of a stochastic source. We defined the search problem for multiple agents in terms of infotaxis [see Eq. (1)]. We also showed why synergy gives rise to an advantage in this type of search. We considered two types of sources. We first demonstrated that a source emitting uncorrelated particles will afford no opportunity for synergy (see Section 3.1). In a search for a Poisson source, multiple coordinated searchers (ones that consider sets of joint moves rather than each considering an independent move) can not hope to do better than multiple independent searchers. Next we showed that, for a source emitting particles with (angular) correlations (see Section 3.2), only synergy or independence is possible (see Fig. 1). The ability of the searchers to leverage synergy depends strongly on their ability to estimate with some accuracy the probability distribution of source locations. These general considerations are crucial for the exploitation of social computation in terms of the design of optimal collective algorithms in particular applications. The next step to making this approach applicable to a broader class of problems, including those not limited to spatial searches, is to generalize the results to more than 2 searchers and to explore how synergy may be best leveraged to give increases in search speed and efficiency.

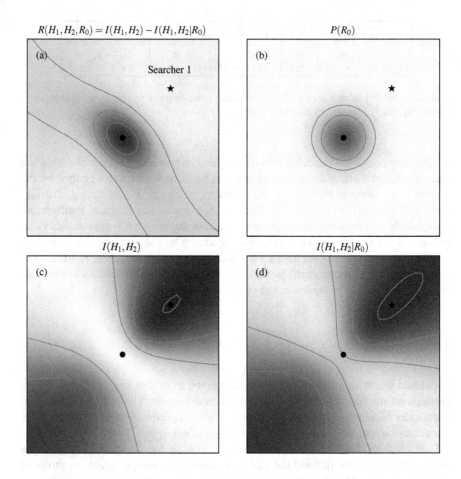

**Fig. 1** Synergy for the two searcher problem with angular correlations. (a) $R(H_1, H_2, R_0)$ as a function of the position of searcher 2 ($r_2$) for a fixed location of searcher 1 ($r_1$, shown as a black star). The most probable source location is in the center (black dot). The white to blue scale indicates $R = 0$ to $R = -2 \times 10^{-5}$ and we note that $R \leq 0$ everywhere. The darker color indicates stronger synergy values when searcher 2 is near the source. The synergy is less when searcher 2 is away from or on the opposite side ($R \approx 0$) of the source. (b) The probability distribution of source locations, peaked at the center: $P(\mathbf{r_0}) = A \exp\left(-|\mathbf{r_0}|^2 / 0.02\right)$, where $A = 1 / \sum_{\mathbf{r_0}} P(\mathbf{r_0})$ is a normalization factor. White to blue indicates $P = 0$ to $P = 0.02$. (c) $I(H_1, H_2)$; (d) $I(H_1, H_2 | R_0)$. In (c) and (d) white to blue indicates $I = 0$ to $I = 6 \times 10^{-5}$. Contour lines have been added to guide the eye. In all frames the data is plotted in a two-dimensional spatial domain of $x, y = [-0.5, 0.5]$ and all vectors are measured from the origin $x = 0, y = 0$. The parameter $\sigma^2 = 1.1$ in Eq. 5.

# References

1. F. Bourgault, T. Furukawa, and H.F. Durrant-Whyte. Coordinated decentralized search for a lost target in a Bayesian world. *Intelligent Robots and Systems, 2003. (IROS 2003). Proceedings. 2003 IEEE/RSJ International Conference on*, 1:48, Oct. 2003.

2. M. Vergassola, E. Villermaux, and B. I. Shraiman. "Infotaxis" as a strategy for searching without gradients. *Nature*, 445:406, 2007.

3. R. S. Sulton and A. G. Barto. *Reinforcement learning: an introduction*. MIT Press, Cambrigde MA, 1998.

4. Iain D. Couzin, Jens Krause, Nigel R. Franks, and Simon A. Levin. Effective leadership and decision-making in animal groups on the move. *Nature*, 433:513, 2005.

5. E. Bonabeau and G. Theraulaz. Swarm smarts. *Scientific American*, pages 72–79, Mar. 2000.

6. H. S. Seung, M. Opper, and H. Sompolinsky. Query by committee. In *COLT '92: Proceedings of the fifth annual workshop on Computational learning theory*, page 287, 1992.

7. Y. Freund, E. Shamir, and N. Tishby. Selective sampling using the query by committee algorithm. In *Machine Learning*, page 133, 1997.

8. S. Fine, R. Gilad-Bachrach, and E. Shamir. Query by committee, linear separation and random walks. *Theor. Comput. Sci.*, 284(1):25, 2002.

9. Luis M. A. Bettencourt, Greg J. Stephens, Michael I. Ham, and Guenter W. Gross. Functional structure of cortical neuronal networks grown in vitro. *Phys. Rev. E*, 75:021915, 2007.

10. L. M. A. Bettencourt, V. Gintautas, and M. I. Ham. Identification of functional information subgraphs in complex networks. *Phys. Rev. Lett.*, 100:238701, 2008.

11. T. M. Cover and J. A. Thomas. *Elements of Information Theory*. Wiley, New York, 1991.

12. E. Schneidman, W. Bialek, and M. J. Berry II. Synergy, redundancy, and independence in population codes. *J. Neurosci.*, 23:11539, 2003.

# The Dilemma of Social Order in Iraq

Michael Hechter[†] and Nika Kabiri[*]

[†], Michael.Hechter@asu.edu, Arizona State University, Tempe, AZ
[*] nkabiri@u.washington.edu, University of Washington, Seattle, WA

Social disorder seems endemic in Iraq. Providing order in this turbulent country is of utmost priority, and many believe that the solution lies in the establishment of a centralized Iraqi state. Some analysts claim that an Iraqi federation would unravel: it would "result in mass population movements, collapse of the Iraqi security forces, strengthening of militias, ethnic cleansing, destabilization of neighboring states, or attempts by neighboring states to dominate Iraqi regions (Baker et al. 2006, 39)." Hence regional leaders should be induced to bolster an independent central government.

This strategy is not much different from those of the recent and more distant past. Paul Bremer's administration (the Coalition Provisional Authority, or CPA) was a highly centralized operation that was later criticized for fueling sectarian violence (Diamond 2004, 3). The constitutionally-based Iraqi government, limited in economic and peace-keeping resources, has been unable to foster peace on its own. Yet lessons from history might have given Bremer and others pause. The British faced a similar dilemma in 1918 after taking control of the territory now known as Iraq. They charged Arnold Wilson, Acting Civil Commissioner of Iraq, with a mandate to provide order in a culturally heterogeneous and well-armed society that had for years been under Ottoman rule. Wilson was faced with a choice of strengthening British direct rule or ruling indirectly through tribal leaders (Wilson 1931). He chose the first option. Not long afterward, massive rebellion across the country forced him out of his job.

Wilson's question – is social order in culturally diverse countries best attained by direct or indirect rule? – is even more pressing today, as civil war, state failure, and terrorism arehave become increasingly part of the global political landscape, For rulers intent on attaining order, the choice of a system of governance is not easy; both answers are plausible. Under indirect rule the high costs of social control are partially borne by subgroups in addition to the state. But indirect rule also has a downside: ceding too much control to local groups can threaten a country's unity. After all, federation has been associated with the fragmentation of the Soviet Union and Yugoslavia (Beissinger 2002; Bunce 1999; Roeder 1991; Woodward 1995). Although centralized rule can offset threats of fragmentation, it offers no guarantee of order either. Some stateless societies (like traditional tribal societies in the Arabian peninsula) manifest a good deal of social order, while direct rule regimes (like Duvalier's Haiti) are often visited by disorder (Chehabi and Linz 1998).

H. Liu et al. (eds.), *Social Computing and Behavioral Modeling*,
DOI: 10.1007/978-1-4419-0056-2_14, © Springer Science + Business Media, LLC 2009

Avoiding indirect rule as a path toward peace in Iraq may be a good policy. But the relationship between types of governance and social order is complex and warrants close scrutiny. Under some conditions, direct rule can instigate resistance to the state, thereby fostering disorder. Under others, indirect rule can discourage unity and hinder order. Since governance structures are pivotal for attaining social order this paper explores their general effects.[1]

## 2 Forms of Governance and Social Order in Iraq

To the degree that a society is ordered, its individual members behave both predictably and cooperatively. Mere predictability is an insufficient condition for social order. The denizens of the state of nature are quite able to predict that everyone will engage in force and fraud whenever it suits them. Hence they are accustomed to taking the appropriate defensive – and offensive – measures. But none of the fruits of social and economic development can occur absent social order. Thus, in a viable social order, individuals must not only act in a mutually predictable fashion; they must also comply with socially-encompassing norms and laws -- rules that permit and promote cooperation (Hechter and Horne 2009).

Social order is not a constant but a variable; it is manifest to the degree that individuals in a territory are free from crime, physical injury, and arbitrary justice. Perfect order is an ideal, so it cannot be attained in Iraq or anywhere else. By any measure, present-day Iraq falls far short of this ideal. Despite the gains made by the recent surge in American troops, Iraqis continue face gunfire, kidnapping, murder, and bombings. Of course, there is a greater amount of order in some Iraqi regions, such as Kurdistan, than others.

How can social order be strengthened? This question has dogged social theorists at least since ancient times. The most popular solution dates from the seventeenth century (Hobbes [1651] 1996): it suggests that social order is the product of direct rule, a multidimensional variable composed of at least two independent dimensions: scope and penetration (Hechter 2004)..The scope of a state refers to the quantity and quality of the collective goods it provides. Welfare benefits, government jobs, state-sponsored schools and hospitals, and a functioning system of justice are examples of such goods. Socialist states have the highest scope; neo-liberal ones the lowest. Scope induces dependence: where state scope is high, individuals depend primarily on the state for access to collective goods.

In contrast, penetration refers to the central state's control capacity – the proportion of laws and policies enacted and enforced by central as against regional or local deci-

---

[1] A more elaborate version of this argument can be found in Hechter and Kabiri (2008).

sion-makers. Penetration is at a maximum in police states in which central rulers seek to monitor and control all subjects within their domain. Polities relying on local agents to exercise control have lower penetration. Scope and penetration often co-vary, but not always. For instance, federal states with similar scope have less penetration than unitary ones.

The relationship between direct rule and social order is contentious. On one view, high scope and penetration foster order by instituting a common culture that provides the shared concepts, values, and norms – thus, the common knowledge (Chwe 2001) – required for cooperation to emerge and persist. Intuitively, cultural homogeneity would appear to be essential for social order. However, the stability of culturally het-erogeneous societies that adopted indirect rule – such as Switzerland, the United Kingdom, and Finland – suggests otherwise. On another view, social order rests not on cognitive commonality, but instead on the authority of central rulers. Indeed, the popu-lar concept of state failure implies the absence of this central authority.

The effectiveness of direct rule was evident in pre-invasion Iraq. Saddam Hussein's rule was highly sultanistic. Fearful that his rule would end the same way it began – by a coup – he went to great lengths to be informed of all threats to his power. He restruc-tured the state's security apparatus monitor citizens with an extensive spy network. Often with little more evidence than a spy's finger-pointing, Saddam would torture and kill suspected Iraqi subversives on a regular basis. Repression kept dissent to a minimum (Roberts 2000; Makya 1998). Saddam's centralized rule did not exclusively rely on penetration, however. Oil revenues allowed him to increase state employment, the size of the military, and the quantity of state-provided welfare benefits: by 1990-1991, 30% of all Iraqi households were dependent on government payments. He also invested heavily in schools, hospitals, food subsidies, and housing projects (Dodge 2003, 160; Tripp 2000, 314). More Iraqis depended on the state for collective goods during Saddam's rule than at any other period in Iraqi history. Saddam's state touched nearly every part of Iraqi life.

Although rational rulers strive for direct rule because ostensibly it maximizes their power, it entails two distinct liabilities. First, direct rule often spurs the opposition of traditional rulers, whose power is threatened as the state advances, and their depend-ents. Second, it is expensive, for direct rulers must assume the costs of policing while providing the bulk of society's collective goods. The British faced both these challenges when they assumed power in Iraq after World War I. They replaced exist-ing Ottoman governing institutions and officials with their own (Tripp 2000, 37). Though positions of power were largely held by aliens, Iraqis were hired to build roads, and railway and irrigation systems (Atiyyah 1973, 214, 219, 224). The British also funded education and medical services to a greater degree than the Ottomans – spending for these services increased threefold from 1915 to 1918.

What was the Iraqi reaction to British direct rule? Local rulers, for one, were threat-ened by British penetration. In some places, protests and revolts appeared almost immediately. As early as 1918 tribal and religious leaders joined forces to challenge British authority (Tripp 2000, 33)., Frustrated by their loss of status under direct rule, Shi'is and Sunnis joined in protest, couching their frustration in terms of Arab self-determination (Yaphe 2003). In the mid-Euphrates region, religious and tribal leaders worked together to formulate strategies for Iraqi independence (Tripp 2000, 41). Secret organizations and parties also emerged.

In April 1920, anti-British sentiment culminated in outright revolt (Dodge 2003, 5). Approximately 130,000 Iraqis rebelled against British rule; it took the British until October to finally quell the violence. They learned a hard lesson. Although appealing in some obvious ways, direct rule has a dark side. Unwilling to deal any longer with the threat of revolt, the British changed tack. Not only were Iraqis allowed to participate in British institutions – Iraqi officials once again governed the countryside – but the Ottoman institutions to which Iraqis had grown accustomed were revived. In 1921, the British installed the Hashemite Amir Faisal as King, marking the beginning of a period of indirect rule in Iraq that lasted thirty-seven years under three different Hashemite monarchs. To curb the power of local Iraqi authorities who might threaten the state's authority, the British used the Royal Air Force (RAF), bombing or threatening to bomb rebellious tribes (Dodge 2003, 154; Sluglett 2003, 7). These new policies meant the British could rule Iraq on the cheap: in 1921, the annual military budget for Iraqi operations was reduced from £25 million to £4 million (Mathewson 2003, 57). The British ultimately chose a governance strategy that was less costly and more popular than direct rule.

In addition to being costly and unpopular, direct rule is problematic for a third reason. The idea that social order is produced in a top-down fashion by resourceful central authorities makes it unclear just how power is ever concentrated in the first place. Top-down theorists have little in the way of an answer, save for the (often valid) idea that it is imposed exogenously on fragmented territories by more powerful states. Beyond its inability to account for primary state formation, this answer underestimates the difficulty that modern states have had in attempting to impose order on less developed societies. The nature of this difficulty becomes apparent when we recall that the emergence of the modern bureaucratic state in Western Europe was slow and and arduous (Elias 1993; Ertman 1997; Gorski 2003). Feudal landholders who managed, against all odds, to secure a preponderance of political power were, for a time, invariably overcome by jealous rivals, rapacious invaders, or intrusive agents of the Vatican. In consequence, the concentration of power oscillated around a highly decentralized equilibrium. This equilibrium persisted for centuries until new military, communications, and industrial technologies allowed power to be concentrated in the modern centralized state.

As has been seen, an increase in state scope and penetration can have perverse effects. Direct rule can fuel the mobilization of both traditional and new groups that carry potential threats to order (as well as the state). When the state extends its control apparatus, it infringes on the traditional self-determination of social groups, particularly culturally distinctive ones. This imposition of a single set of norms on a culturally-diverse population may motivate the leaders of disfavored groups to oppose the state. When a state extends its scope – when it becomes the primary provider of collective goods – it increases individuals' dependence on central rulers.

Yet state-provided collective goods – like education, welfare benefits, and government jobs – are costly to produce and limited in supply. Not everyone receives as much as they wish, and not everyone gets an equal share. Direct rulers become the principal target of redistributive demands by new or traditional groups that can disrupt the social order. Moreover, to the degree that state-provided goods are culturally specific, they are likely to dissatisfy groups that have distinctive preferences regarding such goods.

Consider the recent shift from class- to culturally based politics in advanced capitalist societies (Hechter 2004). By providing the bulk of collective goods in society, the direct-rule state reduces dependence on class-based groups (such as trade unions), thereby weakening them. To the degree that state-provided goods are distributed unequally to individuals on the basis of cultural distinctions, however, the legitimacy of the state in the eyes of these constituents is challenged. This provides such individuals with an incentive to mobilize on the basis of factors such as race, ethnicity, and religion. As a result, social disorder may increase. The rise of nationalist violence has been attributed, in part, to just this mechanism (Hechter 2000).

Saddam's direct rule fostered order and ultimately deflected efforts to topple his regime. But this does not mean that his rule was popular. His generosity in providing state employment opportunities, building schools, roads, and hospitals, and in providing other benefits, was directed primarily at the minority (20%) Sunni Arab population.   This left Shi'i and Kurds as an aggrieved majority. Saddam faced challenges from primarily two directions: the Kurds in the north and the Shi'i in the south. In both these cases local resistance was fed by differential treatment and a lack of concern for regional cultural preferences.

When the regime reneged on its promise of Kurdish autonomy, tensions with the Kurds increased. Saddam's regime made speaking Kurdish in public a crime; in official documents, Kurds were to use Arab names and claim an Arab identity (Human Rights Watch 1995). Although much of Saddam's revenues came from oil production in Kurdish Kirkuk, the Kurds were not given their fair share of the funds. Moreover, Saddam devoted relatively little state revenue to improving Kurdish roads, buildings, or services, nor did the state hire nearly as many Kurds as it did Arabs. Left in a position of relative deprivation, the Kurds rebelled. Sometimes, their revolts were bolstered by Iranian support. At other times, the Iranian government withdrew support from the Kurdish cause. Even so, the Kurds persisted in challenging the regime. Differential treatment can also explain Shi'i resistance. The Shi'i were woefully underrepresented in Saddam's state apparatus. Saddam's secularism also led to Shi'i protests. A cycle of violence ensued (Cockburn and Cockburn 2002).

Ultimately, Saddam managed to quell both Kurdish and Shi'i opposition, but for unusual reasons. Unlike many direct rulers, Saddam had access to large oil revenues and foreign aid which enabled him to avoid many of direct rule's perverse effects. In Kurdistan, Saddam used divide and rule to weaken the two major Kurdish parties. The United Nations' sanctions on Iraq after the 1991 Gulf War were designed to weaken Saddam; likewise, the establishment of no-fly zones was meant to keep Saddam from harming his own people. Ironically, these policies allowed Saddam to increase direct rule over the remaining regions under his control. Direct rule under less-than-ideal circumstances is risky. A state relying on high scope and penetration for rule is likely to face unanticipated forms of resistance.

Effective governance need not reside exclusively with central rulers, however. Under indirect rule, authority is distributed among a number of sub-units or social groups. Distributed authority is especially likely to be effective in culturally-heterogeneous societies. Indirect rulers delegate substantial powers of governance to traditional authorities in return for the promise of tribute, revenue, and military service. Although both direct and indirect rule foster dependence on the state, direct rule results in individuals' dependence, whereas indirect rule entails the dependence of groups.

Indirect rule also has its liabilities, however. Its reliance on solidary groups is only justifiable if these groups do not set out to subvert order or threaten the state. Often solidary groups do subvert social order, however. Consider the large literature on failed states (Kohli 2002), which attributes disorder to a variety of solidary groups that act as hindrances to, and substitutes for, central authority. Moreover, such groups need not be perennially subversive; they can sustain social order at one time and subvert it at another.

Iraqi Kurds are a good example of this. While ruling most of Iraq directly, the British chose indirect rule for Kurdistan. Shaykh Mahmud Barzinji, believed to be influential among the Kurds, was granted governorship of Lower Kurdistan. He was an indirect ruler, freeing the British from the costs of governance in his region. Indirect rule would only have worked if Shaykh Mahmud remained loyal to the British, but he was not. In fact, his influence was far greater than the British imagined.. In May 1919, he declared himself leader of an independent Kurdistan – to the displeasure of other Kurdish Shaykhs who revolted in protest. Revolts ensued, forcing the British to send their military into Kurdistan and reclaim control (Tripp 2000, 34). Indirect rule proved to be disastrous.

When General 'Abd al-Karim Qasim took control in 1958, he provided the Kurdish population with its own cultural space: Kurds were awarded positions in government, opportunities in education, and even some cultural rights. For a brief time Kurds and Arabs coexisted peacefully. The major Kurdish political party at the time publicly recognized Qasim for acknowledging Kurdish cultural rights. By 1959 the Kurds capitalized on their increased autonomy to make greater claims of cultural distinctiveness. Kurdish relations with Qasim remained positive, and a Kurdo-Arab state seemed plausible. Indirect rule in this case proved viable (Natali 2001, 267-268). It was only when Qasim, influenced by events in other regions of Iraq (such as clashes with the country's communist party) and by foreign powers (such as the United States), used force to oppress the Kurds that disorder once again erupted. State arrests and bombings led to Kurds resistance and a renewed energy to challenge state authority (Natali 2001, 269).

The factors that make groups like the Kurds subversive in one context and not in another are not easy to ascertain. It is likely that the Kurds under Qasim were more solidary; the tribal fragmentation in Kurdish Iraq during British rule could have explained why such a governance strategy did not work. It is also likely that the varying degrees of indirect rule may have explained the variation in social order: Qasim provided Kurds with cultural self-determination but he also governed all of Iraq much more directly than the monarchs before him. British indirect rule of Kurdistan may have been more lax, to the extent that it could have played a significant role in the disorder than ensued. In any case, indirect rule is no less risky than direct rule, and must be chosen and designed with care.

## 2 Implications for Social Order in Present-day Iraq

Since each form of rule has strengths and liabilities, choosing an optimal form of governance is a challenge. Direct rule might lead to control of areas known for insurgent activity, but not in all cases; sometimes an increase in scope and penetration can spur violent reaction by local groups. Indirect rule appeases local groups and their desire for autonomy, making them less likely to challenge the state; however, threats under this governance structure come when local leaders use their autonomy to break away from the state altogether. No one governance structure is universally optimal for attaining order.

Full-scale direct rule is a surer means of attaining social order in culturally diverse societies than indirect rule. But since direct rule results in a shift in dependence – for jobs, security, insurance, education, and other collective goods – from traditional authorities and intermediate social groups to the central state, it is extremely costly to implement. The center has but three means of providing the requisite largesse. First, it can do so by its capacity to generate revenue and collective goods endogenously on the basis of robust economic development. This is difficult to accomplish in less developed countries (and no option in the near term for Iraq), but the examples of the four Asian tigers and Market-Leninist China reveal that it is not impossible. A second endogenous means of doing so is through central control over the revenues provided by the export of key resources, like oil. Were it not for Iraqi oil wealth, it is highly unlikely that Saddam would have been effective in implementing direct rule. Absent these means, direct rulers must rely on exogenous sources of aid.

In addition to its manifest costs, direct rule can stir opposition. Competition over collective goods and resistance to encroachments on autonomy can result in challenges to state hegemony by ethnic, religious, or tribal groups. In response to British direct rule, for example, new political parties emerged in Iraq, Sunni and Shi'i groups collaborated, and traditional tribal affiliations were strengthened. Extreme direct rule, as occurred under Saddam, was more effective because it combined extensive welfare benefits with the harshest of sanctions for noncompliance.

Although indirect rule imposes considerably fewer costs on central authorities, it too is costly. In addition to agency costs, which substantially cut into potential central government revenues (Kiser 1999), indirect rule is only effective when it devolves decision-making to groups that are willing to comply with central authorities. What determines whether a given group will be compliant? This question is akin to the classic problem of federalism (Riker 1964), and the solution resides in the center's ability to render the groups (and sub-units) dependent on it for access to vital resources. To the degree that groups are dependent on the center, their leaders' interests will be aligned with those of the state, and they will be motivated to curb their members' oppositional proclivities. This dependence derives from, but is not limited to, financial, kinship, military, and welfare relations with the center. For example, the efficacy of British indirect rule during the first Hashemite monarchy hinged on the RAF's ability to subdue subversive elements in Iraq.

What implications does this analysis have for attaining order in present-day Iraq? Direct rule has proven effective in the past. But unless the government is willing to take Saddam's lead and engage in violent oppression (such as systematic torture) of its people, the strategy is likely to fail. Moreover, the costs of direct rule are high. The Bush administration's military and economic commitment to the occupation was insufficient at the start: enough troops were sent into Iraq to topple Saddam, but not enough to keep the peace (Ricks 2006; Bremer 2006; Diamond 2005). Lacking substantial oil revenues, the Iraqi economy is too frail to support direct rule. Iraq's cultural and ethnic diversity makes agreement about the division of these revenues problematic.

Indirect rule has been effective in Iraq, but not when the state systematically treated cultural groups differently. Ethnic favoritism might be an attractive option, particularly when it involves the weakening of insurgent groups, but indirect rule is most successful when it is fair. Otherwise, competing groups are liable to protest when their constituents are denied resources that are offered to others.

Had the U.S. occupiers entertained some kind of federation, some disorder might have been averted (Galbraith 2006). The CPA did not adopt this strategy because it feared local leaders would have an opportunity to disrupt the nascent state (Bremer

2006). History also provides a warning: British reliance on an indirect ruler to govern Kurdistan led to widespread revolts. But indirect rule has also been viable before in Iraqi history: when Qasim decreased penetration of Kurdish regions, revolts did not ensue. Some form of indirect rule can also work today so long as local groups are dependent on the center and none is perpetually disadvantaged. A three-state solution is necessarily the answer, however. The Iraq Study Group may be correct in saying that the devolution of Iraq into three semi-autonomous regions might threaten order. Instead, the Iraqi state (and the CPA before it) could rely on legitimate leaders of intermediate groups to control their own members, reducing the cost of state governance and local-level resistance (Diamond 2004).

These groups need not reflect the major ethnic and social divisions in Iraqi society today. Rather, what matters is that these groups are among the most solidary in each region, otherwise their indirect rulers could not effectively control them. Unfortunately, some of the most solidary local Iraqi groups have posed the greatest threat to the country's stability. Muqtada al-Sadr's Mahdi Army and al-Qaeda's "Organization in Mesopotamia" are examples. It would prove unwise to offer resources and trust to such organizations given their insurgent and militant tendencies. But not all sectarian groups pose such a threat. The Grand Ayatollah Ali al-Sistani has the support of many of the country's Shi'i and is opposed to the use of violence (Baker et al. 2006). The Iraqi regime can enlist the support of local leaders like Sistani who reject militancy, and in doing so it can manage large sections of the local population while earning their loyalty. The Iraqi state need not rely only local groups that reflect the current sectarian and ethnic divisions in Iraqi politics today. Tribes, though weakened politically during Saddam's regime, are still socially intact (Baram 1997). With enough support, tribal leaders in some regions of the country could prove suitable indirect rules. Where they are influential, tribes have already assisted in keeping peace. Bolstering their strength in sectarian strongholds might offset the growing influence of violent sectarian organizations (Peterson 2008; Michaels 2007). The Iraqi government must move fast, however: as the violence persists, the influence of peaceful local leaders and figureheads like Sistani wanes (Baker et al. 2006).

To accomplish these goals on the cheap is dubious at best; moreover, it cannot occur overnight. In the meantime, an increasingly vigorous resistance consumes resources that could otherwise be used for vitally important civil investment. Insurgent groups are targeting each other today so as to shore up their political position for when, sooner or later, the Americans go home. These events hark back to the 1920 revolt against British rule.

# References

1. Attiyah, Ghassan R. 1973. Iraq, 1908-1921: A Socio-political Study. Beirut: Arab Institute for Research and Publication.
2. Baker, James A. II, Lee H. Hamilton, Lawrence S. Eagleburger, Vernon E. Jordan, Jr., Edwin Meese III, Sandra Day O'Connor, Leon E. Panetta, William J. Perry, Charles S. Robb, and Alan K. Simpson. 2006. The Iraq Study Group Report. New York: Vintage Books.
3. Baram, Amatzia. 1997. "Neo-tribalism in Iraq: Saddam Hussein's Tribal Policies 1991-96." International Journal of Middle East Studies 29: 1-31.
4. Beissinger, Mark R. 2002. Nationalist Mobilization and the Collapse of the Soviet State. Cambridge, U.K.; New York: Cambridge University Press.
5. Bremer, Paul. 2006. My Year in Iraq. New York: Simon and Schuster.
6. Bunce, Valerie. 1999. Subversive Institutions: The Design and the Destruction of Socialism and the State. New York: Cambridge University Press.
7. Chehabi, H. E., and Juan J. Linz. 1998. Sultanistic Regimes. Baltimore: Johns Hopkins University Press.
8. Chwe, Michael Suk-Young. 2001. Rational Ritual: Culture, Coordination, and Common Knowledge. Princeton, N.J. ; Oxford: Princeton University Press.
9. Cockburn, Andrew, and Patrick Cockburn. 2002. "Saddam at the Abyss." In Inside Iraq, ed. John Miller and Aaron Kennedy. New York: Marlowe and Company, Pp. 167-207.
10. Diamond, Larry. 2004. "Testimony to the Senate Foreign Relations Committee." <http://www.stanford.edu/~ldiamond/iraq/Senate_testimony_051904.htm> (October 7, 2006).
11. Diamond, Larry. 2005. Squandered Victory. New York: Henry Holt and Company.
12. Dodge, Toby. 2003. Inventing Iraq. New York: Columbia University Press.
13. Elias, Norbert. 1993. The Civilizing Process. Oxford [England]; Cambridge, Mass.: Blackwell.
14. Ertman, Thomas. 1997. Birth of the Leviathan: Building States and Regimes in Medieval and Early Modern Europe. Cambridge: Cambridge University Press.
15. Galbraith, Peter W. 2006. The End of Iraq: How American Incompetence Created a War Without End. New York: Simon & Schuster.
16. Gorski, Phillip. 2003. The Disciplinary Revolution. Chicago: University of Chicago Press.
17. Granovetter, Mark. 1973. "The Strength of Weak Ties." American Journal of Sociology 78(May):1360-1380.
18. Hechter, Michael. 2000. Containing Nationalism. Oxford: Oxford University Press.
19. Hechter, Michael. 2004. "From Class to Culture." American Journal of Sociology 110 (September):400-445.
20. Hechter, Michael and Christine Horne. 2009. Theories of Social Order: A Reader. Stanford: Stanford University Press.
21. Hechter, Michael and Nika Kabiri. 2008. "Attaining Social Order in Iraq." In Order, Conflict and Violence, eds. Stathis Kalyvas, Ian Shapiro, and Tarek Masoud. Cambridge: Cambridge University Press. Pp. 43-74.
22. Hobbes, Thomas. [1651] 1996. Leviathan. Oxford ; New York: Oxford University Press.
23. Human Rights Watch/Middle East. 1995. Iraq's Crime of Genocide: The Anfal Campaign Against the Kurds. New Haven: Yale University Press.
24. Kiser, Edgar. 1999. "Comparing Varieties of Agency Theory in Economics, Political Science, and Sociology: An Illustration from State Policy Implementation." Sociological Theory 17:146-179.
25. Kohli, Atul. 2002. "State, Society and Development." In Political Science: State of the Discipline, eds. Ira Katznelson and Helen V. Milner. New York: W. W. Norton; Washington, DC: American Political Science Association. Pp. 84-117.
26. Makiya, Kanan. 1998. Republic of Fear. Berkeley: University of California Press.

27. Mathewson, Eric. 2003. "Assessing the British Military Occupation." In U.S. Policy in Post-Saddam Iraq, eds. Michael Eisenstadt and Eric Mathewson. Washington DC: Washington Institute for Near East Policy. Pp. 52-66.
28. Michaels, Jim. 2007. "U.S. gamble on sheiks is paying off – so far; Military's alliances with tribes in Iraq foster peace." USA Today, December 27, Thursday, News p. 1A.
29. Natali, Denise. 2001. "Manufacturing Identity and Managing Kurds in Iraq," In Right-sizing the State, eds. Brendan O'Leary, Ian S. Lustick, and Thomas Callaghy. Oxford: Oxford University Press. Pp. 253-288.
30. Peterson, Scott. 2008. "As violence drops, Iraqi tribes begin to make amends." Christian Science Monitor, October 9, Thursday, World, p. 1
31. Ricks, Thomas E. 2006. Fiasco: The American Military Adventure in Iraq. New York: Penguin Press.
32. Riker, William H. 1964. Federalism: Origin, Operation, Significance. Boston: Little-Brown.
33. Roberts, Paul William. 2000. "Saddam's Inferno." In Inside Iraq: The History, the People, and the Modern Conflicts of the World's Least Understood Land, eds. John Miller and Aaron Kenedi. New York: Marlowe & Company. Pp. 101-124.
34. Roeder, Philip G. 1991. "Soviet Federalism and Ethnic Mobilization." World Politics 43(January):196-232.
35. Sluglett, Peter. 2003. "The British Legacy." In U.S. Policy in Post-Saddam Iraq: Lessons from the British Experience, eds. Michael Eisenstadt and Eric Mathewson.
36. Tripp, Charles. 2000. A History of Iraq. Cambridge: Cambridge University Press.
37. Weber, Max. [1919-1920] 1948. "The Protestant Sects and the Spirit of Capitalism." In From Max Weber: Essays in Sociology, eds. Hans Gerth and C. Wright Mills. New York: Oxford University Press. Pp. 302-322.
38. Wilson, A. T. 1931. Loyalties, Mesopotamia, vol. II, 1917-1920: A Personal and Historical Record. London: Oxford University Press.
39. Woodward, Susan L. 1995. Balkan Tragedy : Chaos and Dissolution After the Cold War. Washington, D.C.: Brookings Institution.
40. Yaphe, Judith. 2003. "The Challenge of Nation Building in Iraq." In U.S. Policy in Post-Saddam Iraq, eds. Michael Eisenstadt and Eric Mathewson. Washington DC: Washington Institute for Near East Policy. Pp. 38-51.

# A Socio-Technical Approach to Understanding Perceptions of Trustworthiness in Virtual Organizations

Shuyuan Mary Ho[†]

[†] smho@syr.edu, School of Information Studies, Syracuse University, Syracuse, New York

**Abstract** This study adopts a socio-technical approach to examining perceptions of human *trustworthiness* as a key component for countering insider threats in a virtual collaborative context. The term *insider threat* refers to situations where a critical member of an organization behaves against the interests of that organization, in an illegal and/or unethical manner. Identifying and detecting how this individual's behavior changes over time - and how anomalous behavior can be detected - are important elements in the preventive control of insider threat behaviors. This study focuses on understanding how anomalous behavior can be detected by observers. While human observations are fallible, this study adopts the concept of human-observed changes in behavior as being analogous to "sensors" on a computer network. Using online team-based game-playing, this study seeks to re-create realistic situations in which human sensors have the opportunity to observe changes in the behavior of a focal individual – in the case of this research, a team leader. Four sets of experimental situations are created to test the core hypotheses. Results of this study may lead to the development of semi-automated or fully-automated behavioral detection system that attempts to predict the occurrence of malfeasance.

## 1 Introduction

Trust in any organization is critical because it enables individuals the ability to collaborate with one another. This is even true in a virtual organization (VO) [4]. However, trust can be misused by critical team members for purposes of unethical or illegal behavior. According to 2007 CSI Survey, financial losses caused by computer crimes soared to $67 million in 2007, up from $52.5 million in 2006 [16]. Among those losses, nearly 37 percent of respondents attributed more than 20 percent of losses to be caused by insiders. This indicates an increase of insider abuse within network resources, from 42 to 59 percent compared with the 2006 CSI/FBI Survey. Insider misuse of authorized privileges or abuse of network access has caused significant damage and loss to corporate internal information assets. While employees are essential to the productive operation of an organization, their inside knowledge of corporate resources can also threaten corporate security, as a result of improper use of information resources. Such improper uses are often

H. Liu et al. (eds.), *Social Computing and Behavioral Modeling*,
DOI: 10.1007/978-1-4419-0056-2_15, © Springer Science + Business Media, LLC 2009

termed by security experts as "insider threats," where trust is betrayed within an organization. The capability to anticipate the occurrence of improper uses of corporate internal resources within some acceptable range is valuable. Thus, being able to know a person's trustworthiness will enable a virtual organization to achieve its goals and maintain productivity.

This research explores basic mechanisms on how to detect changes in the trustworthiness of an individual who holds a key position in an organization through the observation of overt behavior – including communication behavior – over time. The class of individual whose behavior is the focus of this investigation is someone who has been granted access to and authority over important information within the organization. This access and authority is what is meant by "holding a key position" in the organization. Employees with access and authority have the most potential to cause damage to that information, to organizational reputation, or to the operational stability of the organization.

This paper contains five major sections describing this study. Since the problem gap of this research is aimed at the insider threat, my research question is raised to understand this phenomenon as stated in the *Problem-based Question*. I then synthesized the theoretical foundation of trustworthiness attribution in the *Theoretical Frameworks*. I described the rationales and settings of a novel research design by recreating social interactions in which issues of trust have raised in the *Method* section. A factorial design is sketched in this section to test my hypotheses. The *Preliminary Discussions* examines the initial findings of both Pilot as well as the Full-Scale Experiments. The *Conclusions and Future Work* section summarizes this research in progress.

## 2 Problem-based Question

The phenomenon of insider threats is a social, human behavioral problem [8]. In the *Insider Threat Study by CERT (2004-2005),* the US DoD[1], DHS[2], and Secret Service investigated various insider threat cases and discovered that embedded in a mesh of communications, a person given high social power but with insufficient trustworthiness can create a single point of trust failure [12, 14]. Thus, "insider threat" as an organizational problem is defined as *a situation where a critical member of an organization with authorized access, high social power and holding a critical job position, inflicts damage within an organization.* In a way, *this critical member behaves against the interests of the organization, generally in an illegal and/or unethical manner.*

This study, based on the above definition, examines basic mechanisms for detecting changes in the trustworthiness of an individual who holds a key position in an organization, by observing overt behavior – including communication behavior

---

[1] US DOD stands for the US. Department of Defense.
[2] DHS stands for the Department of Homeland Security.

– over time. Since Steinke [20] suggests that it is possible to detect cheating behavior without directly observing the individual, the overarching question is: *What changes of behaviors can reflect a downward[3] shift in the trustworthiness of a critical member in a virtual or physical organization which might signal possible insider threats?* My hypothesis is that the downward shift in a person's trustworthiness can be reflected in his or her behavior. And, the inconsistency and unreliability in this actor's unexpected behaviors when compared to his or her communicated intentions can be detected by the observers' subjective perceptions over time. The term "observers" refers to the members of his or her close social network.

## 3 Theoretical Frameworks

This framework integrates trust theories and attribution theory. The purpose of this framework is to explain and predict whether the trustworthiness of a person who has social power and holds a key position, can be perceived by members of his social networks[4] through their attribution of the target[5]'s behavior compared with his or her communicated intentions. Rotter [17] defined trust as a generalized expectancy - held by an individual or a group - that the communications of another (individual or group), including word, promise, verbal or written statement, can be relied upon [19] regardless of situational specificity [17]. However, Rotter [18] asserted that "trust and trustworthiness are closely related," but trust depicts a relationship among two or multiple parties or actors while trustworthiness is an attribute or a quality of a person. Trustworthiness[6] is then defined as a generalized expectancy concerning a target person's degree of correspondence between communicated intentions and behavioral outcomes that are observed[7] and evaluated, which remain reliable, ethical and consistent, and any fluctuation between target's intentions and actions does not exceed the observer's expectations over time [5, 11, 19].

Ajzen stated that an individual's intention can be understood by his or her information behavior [1, 2]. Moreover, the quality and trustworthiness can be perceived and assigned with a meaning by a social network in a close relationship [9, 10, 15]. This act of assigning meaning to an individual's behavior [3] within an

---

[3] Same as *unethical*, or *illegal*.

[4] mainly peers, subordinates and associates.

[5] refers to a human participant, or a human subject that is being studied.

[6] Trustworthiness is portrayed and defined as "reliable, dependable, responsible, loyal, honorable, ethical, moral and incorruptible" (Oxford American Dictionary and Thesaurus with Language Guide: *Trustworthy*).

[7] Same as *perceived*.

organization is what is characterized as attribution. Attribution theory has been adopted to understand how people attribute (or assign) the causes of others' behaviors [6, 7]. The attribution of the target's behavior by observers is determined by observers' judgment that the target intentionally or unintentionally [6] behaves in a way that is attributable to either external (situational) causality or internal (dispositional) causality [13]. Because all human beings are of the same species and born with similar types of features and functions, a person should "know" and be able to "sense" from his own perceptions and with his judgment as to how the world operates [6, 13] despite the fact that sometimes those attributions may not be accurate or valid. Moreover, an individual's observed behavior can be interpreted in a single observation – or through multiple observations over time.

The main construct of this framework is human-perceived trustworthiness, particularly regarding those individuals' who hold critical positions in an organization. The intentions of the target subject are reflected in the information behavior [1, 2] through the lenses of perceived trustworthiness in an organization. Figure 1 depicts the theoretical framework for attributing trustworthiness based on observed behavior of the target. A multi-level analysis of mixed "lenses" is adopted in looking at how behavior of a target is influenced by his or her intention within existing organizational norms.

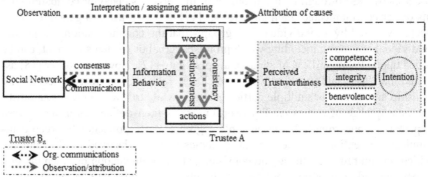

Figure 1: Theoretical Framework

In Figure 1, the relationship of the constructs is represented by arrows. Blue arrows represent communication between the target and the social network. Red arrows represent the observation and attribution by members of the social network regarding the target's behavior. Green arrows represent three attribution principles depicted by Kelley [13]. In this framework, the communications among the target's social network (including communications to and from the target) sheds light on the target's perceived trustworthiness. In other words, members of the social network attribute (or assign) meaning to the target's trustworthiness level based on their observations of the target's behavior.

Trust represents predictability and a correspondence between behavior and words. If we refer an observer as B and a target person as A, unexpected behaviors – actions that do not fit the usual pattern – would trigger B's attributional analysis of A. The following hypothesized theoretical mechanism is at work. With *external attribution* to plausible circumstances, B attributes any anomalous behavior of A to circumstances beyond A's control. As a result, B's perception of A's trustworthiness is not adversely affected (at least with respect to integrity; trust in A's *competence* may be affected, but that is beyond the scope of this study). However with an *internal attribution* by B of A's anomalous behavior, B concludes that A is performing the anomalous behavior by choice, and therefore B may downgrade the perception of A's trustworthiness. The empirically well-established reasons for attributing behavior to external or internal causes (e.g., distinctiveness, consensus, and consistency) apply to B's thinking processes about A's behavior. In summary, it is B's attributions about A's anomalous behavior that influences perceived changes in trustworthiness.

# 4 Method

This study seeks to identify indicators of abnormal behavior and the basic criteria of trustworthiness assessment. The "Leader's Dilemma" game is a series of simulated, controlled "honeypot" situations created to test how a leader's trustworthiness is perceived by his team members. This game simulates a virtual organization including the front-end (overt) where normal business operates while the back-end (covert) information is captured regarding how a target is influenced. The "front-end" data includes emails, chats, discussion blogs, surveys, interviews, etc. for team members within the organization, from which we analyze indicators of the "back-end" manifestations of trust within a virtual organization. A virtual organization refers to a group of individuals whose members and resources may be dispersed geographically, but function as a coherent unit through the use of cyber-infrastructure. This group of individuals is team-based and goal-oriented, where leaders and subordinates work together to achieve pre-determined goals. In these experimental settings, the Game-Master (G) has the role of manipulating the dynamics of the virtual competition. The Experimenter takes on the role of a Moderator (M) in these online games. Team-Leader (A) is the target, who is appointed from among the team participants by the Game-Master. Team members are observers ($B_n$), who work with Team-Leader in achieving their pre-determined goals. Figure 2 depicts how each role in this virtual contest is situated.

A pilot study[8] and a full-scale experiment[9] were simulated based on the above definition of a virtual organization. The stated goal of this experiment was for a team to solve a series of brain teasers, as task assignments, in a given timeframe.

---

[8] Syracuse University IRB#07-276, conducted in Fall 2007.
[9] Syracuse University IRB#07-276, conducted in Fall 2008.

The virtual contest was manipulated in an online game environment[10] when a dishonesty gap was intentionally created by offering "bait" to the Team-Leader. A conflict of interest arises between the Team-Leader and the team members which causes the Team-Leader to face an "ethical dilemma." Bait was offered in a form of a micro-payment system, which connects a monetary value to real-world rewards. A "mole" player is embedded in the team, to question the leader, raise tensions, and stir up discussion in the team. This enhances awareness within team members as to what the leader is doing or thinking. Moreover, peer influence is enhanced by having a third Team-Leader chatting and persuading this target to accept the bait and betray his team. With the awareness of knowing that this is the critical point in determining whether this experiment is successful or not, the bait given to the leader has to be invisible to the teams, and sufficiently tempting that the leader will risk taking the bait.

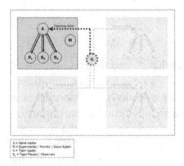

| Treatment / Time / Setting | without treatment/bait; $(B_0)$ | | | with treatment/bait; $(B_1)$ | | |
|---|---|---|---|---|---|---|
| | Time $(T_1)$ | Time $(T_2)$ | Time $(T_3)$ | Time $(T_1)$ | Time $(T_2)$ | Time $(T_3)$ |
| Increase group sensibility $(S_1)$ | Group 1 Average | | | Group 3 Average | | |
| Decrease group sensibility $(S_2)$ | Group 2 Average | | | Group 4 Average | | |

Figure 3: 2×2×3 factorial design

Figure 2: Simulated Game Design

## 4.1 Factorial Design

The same research settings designed in the pilot have been simulated in a full-scale experiment. A 2×2×3 factorial design is developed to generalize the findings. Four sets of simulated case studies of online games are conducted (Figure 3). In other words, 12 sets of group observations are obtained. While the dependent variable (response) is target's perceived trustworthiness, major independent variables (factors) include: the bait ($B_0$ and $B_1$) as the treatment, a mole that increases or decreases group sensibility ($S_1$ and $S_2$) by either encouraging or discouraging conversations about the team-leader, and time ($T_1$, $T_2$ and $T_3$) representing measurement obtained from each day, in particular, after the conflict of interest between the team-leader and the team members is created.

## 4.2 Variables

---

[10] The Learning Management System (http://ischool.syr.edu/learn/) is hosted by School of Information Studies, Syracuse University.

The dependent variable represents the response measure while the independent variables are the factors. The relationship between dependent variable (Y) and independent variable (X) can be expressed as: $Y = f(X_1, X_2, X_3)$, where, the dependent variable is the target's perceived trustworthiness (Y) in terms of taking the bait. The first independent variable (factor 1) is the observers' attribution ($X_1$) in terms of group sensitivity toward target's words (target's communicated intentions). The second independent variable (factor 2) is the observers' attribution ($X_2$) in terms of group sensitivity toward target's actions (target's information behavior). The third independent variable (factor 3) is the Time ($X_3$).

## 4.3 Hypotheses

My main hypothesis is that the downward shift in a person's trustworthiness can be reflected in his or her behavior. And, the inconsistency and unreliability in this actor's unexpected behaviors - when compared to his or her communicated intentions - can be detected by the observers' subjective perceptions over time. There are four hypotheses, which support my main hypothesis (Figure 4).

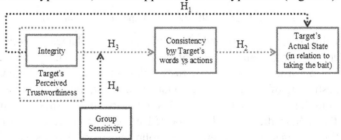

Figure 4: Hypotheses

*Hypothesis 1* ($H_1$): There is a positive relationship between the target's actual state and the group observation of the target's perceived trustworthiness, in terms of his or her integrity. This means that if the target has taken the bait, it can be successfully attributed by the observers over time. If the target has not taken the bait, it will not trigger any suspicion in observers' attribution.

*Hypothesis 2* ($H_2$): When target's actual state is positive (meaning that he has taken the bait), the group can reach consensus about target's inconsistency between communciated intentions (words) and information behavior (actions).

*Hypothesis 3* ($H_3$): When target's perceived trustworthiness is relatively low, observers tend to attribute inconsistency in his or her words and actions.

*Hypothesis 4* ($H_4$): The group sensitivity has a significant influence on the perceived trustworthiness, in terms of integrity, of the target. The higher the group sensitivity is, the more likely group can detect inconsistency between target's words and actions.

## 5 Preliminary Discussions

Activities in Day 1 and Day 2 were designed for the purpose of enhancing group rapport. The bait offered at the end of Day 2 game stirred up a conflict of interest for the target team-leader. I hypothesized that attribution among four groups would differ as illustrated in Figure 5. The results of participant observations confirmed these hypothesized situations during the full-scale experiment.

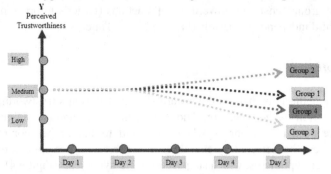

Figure 5: Four Hypothesized and Experimented Situations

Specifically, these experimental situations can be explained as follows. The team-leaders for Group 1 and 2 were not presented with bait. The group sensitivity toward the team-leader's perceived trustworthiness was increased in Group 1 through questioning of the leader's behavior. As a result, the Group 1 team-leader suffered from being questioned about his integrity and fairness by his team members throughout the game. He allocated more micro-payments to his team members than to himself. In Group 2, the group sensitivity was decreased toward the team leader's perceived trustworthiness. The general perceptions of the leader's trustworthiness remained high. The results showed that Group 2 had the most harmonous teamwork atmosphere.

The team-leaders in the pilot study, and also in Groups 3 and 4 in the full-scale experiment were all presented with monetary bait. All three team-leaders took the bait without letting the team members know. As for Group 3, bait was offered to the team-leader and the group sensitivity was enhanced through encouraging discussion about the target's questionable behavior. The attribution toward the perceived trustworthiness dropped in Group 3. The target team-leader in Group 4 was also given, bait but group sensitivity was reduced through discouraging discussion. The attribution toward the target's perceived trustworthiness dropped accordingly. However some team members could sense the team-leader's behavioral change.

In all these experiments, the participants did not know that the target Team-Leader was being manipulated by the Game-Master – which occurred in the background. Since insufficient evidence existed regarding the target, the resulting perceptions depended primarily on whether the target's behavior was generally reliable or ethical, and on the outcome itself. The perception of all targets' behavior

120

was nearly all positive from Day 1 through Day 4 during the experiment. This inferred that the targets' competence in leading the team was found to be satisfactory, and their communicated intention was found to be consistent in terms of his information behavior. In this case, the "anomalous" behavior was not found to be significant. However, the outcome of the targets' behavior for Group 3 and 4 turned negative on Day 5, and the level of the target's integrity dropped as a result of taking the bait. Thus, the target's anomalous behavior was found to be significant.

The swift trust that developed amongst the team players towards the target was caused by the leadership "halo effect" until Day 4. At this point, the target showed signs of dishonesty on a couple of occasions. In my preliminary testing of the hypotheses, there was a corresponding relationship between the target's actual state and the group observation of the target's perceived trustworthiness. When the target takes the bait, the perceptions of his trustworthiness are decreased. Generally the group sensitivity has a significant influence on the target's perceived trustworthiness, in terms of his integrity.

# 6 Conclusions and Future Work

This study explains and has capacity to predict insider threats by attributing trustworthiness of individuals in virtual organizations. However, it obviously has some limitations. Human perception is not fully reliable due to the fact that not all information is made transparent to the perceivers. Humans attribute their perception of people's trustworthiness based on limited social interactions; however, group consensus overcomes the limitations of fallible individual observations. Most of the attributions are context-specific, time-dependent and combined with judgment regarding the target's capability. Basic struggles of personal gain, selfishness and greediness remain. Ethical values and moral standards are vaguely defined by society and therefore vaguely adopted by individuals.

The findings demonstrate that it is possible to trace and detect anomalous information behavior of an insider leader with a high degree of accuracy, although the leader's change in behavior is subtle. It is believed that this framework of trustworthiness attribution can be generalized through future understanding of human conversational acts and logic. Some immediate future work includes using a mixed method to analyze conversations and survey results derived from the rich dataset in order to test my hypotheses. Conversations will be analyzed and extracted from this study to build a trustworthiness detection model, which can be formalized to create a socio-technical system for insider threat prediction.

**Acknowledgements** I thank Jeffrey M. Stanton, my advisor, for his constant support and insight. I thank Conrad Metcalfe for his helpful comments and editing assistance.

# References

1. Ajzen, I. (1991). The Theory of Planned Behavior. *Organizational Behavior and Human Decision Processes*, 50(2), December 1991, 179-211.
2. Beck, L., & Ajzen, I. (1991). Predicting Dishonest Actions Using the Theory of Planned Behavior, *Journal of Research in Personality*, 25(3), September 1991, 285-301.
3. Giddens, A. (1979). *Central Problems in Social Theory: Action, Structure and Contradiction in Social Analysis*. Berkeley, CA: University of California Press.
4. Handy, C. (1995). Trust and the Virtual Organization. *The Harvard Business Review Book Series: Creating Value in the Network Economy*. Harvard Business School Press, Boston, MA, 1999, 107-120.
5. Hardin, R. (2003). Gaming trust. In E. Ostrom & J. Walker (Eds.), *Trust and reciprocity: Interdisciplinary lessons from experimental research* (pp. 80-101). New York: Russell Sage Foundation.
6. Heider, F. (1958). The psychology of interpersonal relations. New York: John Wiley & Sons.
7. Heider, F. (1944). Social perception and phenomenal causality. *Psychological Review*, 51, 358-374.
8. Ho, S. M. (2008). Attribution-based Anomaly-Detection: Trustworthiness in an Online Community. *Social Computing, Behavioral Modeling, and Prediction*. Springer: January 2008, 129-140.
9. Holmes, J. G., and Rempel, J. K. (1989a). "Trust in Close Relationships." In *Review of Personality and Social Psychology*, Vol. 10, ed. C. Hendrick. Beverly Hills, CA: Sage Publications.
10. Holmes, J. G., & Rempel, J. K. (1989b). Trust in close relationship. In C. Hendrick. (Ed.), *Close relationship* (pp. 187-220). Newbury Park: CA: Sage.
11. Hosmer, L. T. (1995). Trust: The Connecting Link between Organizational Theory and Philosophical Ethics. *Academy of Management Review*, 20(2), Apr., 1995, 379-403.
12. Keeney, M., Kowalski, E., Cappelli, D., Moore, A., Shimeall, T., and Rogers, S. (2005). "*Insider Threat Study: Computer System Sabotage in Critical Infrastructure Sectors*." National Threat Assessment Center, U.S. Secret Service, and CERT® Coordination Center/Software Engineering Institute, Carnegie Mellon, May 2005, pp.21-34. Obtained from http://www.cert.org/archive/pdf/insidercross051105.pdf on April 10, 2007.
13. Kelley, H.H. (1973). The Process of Causal Attribution, *American Psychologist*, Feb 1973, 107-128. Obtained from http://faculty.babson.edu/krollag/org_site/soc_psych/kelly_attrib.html on July 5th, 2007.
14. Randazzo, M. R., Keeney, M., Kowalski, E., Cappelli, D., and Moore, A. (2004). Insider Threat Study: Illicit Cyber Activity in the Banking and Finance Sector. National Threat Assessment Center, U.S. Secret Service, and CERT® Coordination Center/Software Engineering Institute, Carnegie Mellon, August 2004. Obtained from http://www.secretservice.gov/ntac/its_report_040820.pdf n April 10, 2007.
15. Rempel, J.K., Holmes, J.G., & Zanba, M.D. (1985). Trust in close relationship. *Journal of Personality and Socio Psychology*, Vol. 49, p. 95-112.
16. Richardson, R. (2007). 2007 *CSI Computer Crime and Security Survey*. Computer Security Institute.
17. Rotter, J. B. (1980). Interpersonal Trust, Trustworthiness, and Gullibility. *American Psychologist*, Jan 1980, 35(1), 1–7.
18. Rotter, J.B. and Stein, D.K. (1971). Public Attitudes Toward the Trustworthiness, Competence, and Altruism of Twenty Selected Occupations. *Journal of Applied Social Psychology*, Dec 1971, 1(4), 334–343.
19. Rotter, J.B. (1967). A new scale for the measurement of interpersonal trust. *Journal of Personality*, 35 (4), 651–665.
20. Steinke, G. D. (1975). The prediction of untrustworthy behavior and the Interpersonal Trust Scale. Unpublished doctoral dissertation, University of Connecticut, 1975.

# Prior-free cost sharing design: group strategyproofness and the worst absolute loss

Ruben Juarez

rubenj@hawaii.edu, Economics Department, University of Hawaii, Honolulu, HI

**Abstract** A service is produced for a set of agents at some cost. The service is binary, each agent either receives service or not. A mechanism elicits the willingness to pay of the agents for getting service, serves a group of agents and covers the cost by charging those agents.

We consider the problem in which the designer of the mechanism has little information about the participants to be able to form a prior-belief about the distribution of their possible types. This paper investigates mechanisms that meet two robust properties that are meaningful in this setting.

On one hand, we look at mechanisms that prevent coordination between any group of agents (henceforth called group strategyproof). On the other hand, we look at mechanisms that are optimal using the worst absolute surplus loss measure.

The mechanisms that are optimal and group strategyproof are characterized for different shapes of cost functions.

## 1 Introduction

A good or service is produced at some non-negative cost. Agents are interested in getting service (or consuming at most one unit of that good) and are characterized by their private monetary valuation for it (which we call their utility). A mechanism elicits these utilities from the agents, serves some agents, and covers the cost by charging only the agents who are served.

These traditional mechanisms find several applications depending on the shape of the cost function. The most discussed problem in the literature is the case of decreasing marginal cost, also referred as economies to scale or natural monopolies. Applications include the production of cars, pharmaceutical goods, software, railways, telecommunications, electricity, water and mail delivery. The canonical example in network economics is the network facility location problem with a single facility, where there is a personalized connection cost of the agents to the facility and a fixed cost of opening the facility.

On the other hand, increasing marginal cost finds applications in the exploitation of natural resources (e.g. oil, natural gas or fisheries). Another interesting application is the scheduling of jobs, where the disutility of the agents is the waiting time until being served (see Cres and Moulin[2001] and Juarez[2008a] for discussions).

H. Liu et al. (eds.), *Social Computing and Behavioral Modeling*,
DOI: 10.1007/978-1-4419-0056-2_16, © Springer Science + Business Media, LLC 2009

The management of queues in networks, for instance the Internet, is the canonical example.

One downside of the traditional analysis of this problem (and more generally of mechanism design) is that it usually assumes the designer of the mechanism has a lot of information. Specifically, that he knows the (probability distribution of the) type of agents that are participating in the game, and based on this distribution the designer chooses the mechanism to implement. For instance, traditional decision theory computes the optimal decision based on the beliefs of the mechanism designer (e.g. Myerson [1981]). However, there is little discussion in the literature on how to form these initial beliefs, assuming the designer can form them at all.

We consider the problem of designing mechanisms in a setting in which the designer does not have enough information about the (potential) participants to be able to form a prior belief about the distribution of their possible types. For instance, it may be that we are dealing with agents in a large network like the Internet, where the variation of the agent types is huge; or that doing a market survey would be costly and time consuming. For these problems, it is usually the case that traditional (Bayesian) techniques do not work well. Therefore, different mechanisms and techniques should be proposed.

This paper looks at two strong properties that are meaningful in the prior-free cost sharing model: group strategyproofness and a worst-case measure. On one hand, group strategyproofness (GSP) rules out coordinated misreports under any possible information context. In particular this property is robust because it works whether the information on individual characteristics is private or not. That is, GSP works whether a small or large group of agents can coordinate misreports.

On the other hand, we use the worst absolute surplus loss measure (*wal*), that is the supremum of the difference between the efficient surplus and the surplus of the mechanism, where the supremum is taken over all utility profiles. This measure has been used recently in the literature as a second-best efficiency measure in similar cost sharing models.[1] This measure is prior-free because it does not depend on the (potential) distribution of valuations.

We present our main results using two illustrative examples. On one hand, the results for decreasing marginal cost are shown in the single facility location problem. On the other, the main result for increasing marginal cost functions is illustrated in the generalized scheduling problem. We discuss some generalization of the results, related literature and open problems in the last section. Due to the short nature of this paper, most technical details are skipped, but can be found in the cited references.

## 2 Single facility location problem

There is a group of agents $N = \{1, \ldots, n\}$ interested in getting connected to the Internet. Every agent $i$ has a private monetary valuation $u_i$ for getting connected.

---

[1] See Moulin and Shenker[2001] and Juarez[2008a] for applications of the *wal*-measure. See Moulin[2008] for an application of the *best relative gain*, a similar worst-case measure.

There is a single facility which can connect them. There is fixed cost $F$ for opening the facility and a personalized connection cost of the agents to the facility. We denote by $c_i$ the cost of agent $i$ for getting connected. Therefore, if the group of agents $S$ gets connected, the total cost of serving them is $C(S) = F + \sum_{i \in S} c_i$. If no agents gets service then there is no connection cost.

A cost-sharing mechanism collects bids from the agents for getting connected, picks a winning set of agents who are going to get the service, and determines the prices for the winners.

Consider the following efficient mechanism. If the bid vector is $(u_1, \ldots, u_n)$, then serve the coalition of agents $S^*$ that maximizes the surplus, that is $S^* = argmax_{S \subseteq N} \sum_{i \in S} u_i - C(S)$. Allocate non-negative payments among the agents in $S^*$ in any way such that every agent $i$, $i \in S^*$, never pays more than his utility $u_i$, and total cost $C(S^*)$ is at least covered by the payments. For instance, payments may be allocated following an arbitrarily order.

**Proposition 1 (Juarez(2008b)).** *Any efficient mechanism is not group strategyproof.*

Therefore, there is a trade off between choosing a mechanism that is GSP and one that is efficient. To see this, consider the following *equal fixed cost mechanism (EFC)*. Initially all agents are offered to be connected to the server at prices given by $p^N = (\frac{F}{n} + c_1, \ldots, \frac{F}{n} + c_n)$. That is, agents $i$ is offered to buy a unit at his connecting price $c_i$ plus an equal share of the fixed cost. If all agents accept the offer (that is $u_i \geq p_i^N$ for all $i \in N$), they all get service at those prices. If only $S$ agents accepted the offer, $S \subsetneq N$, then only those agents are re-offered to get a unit of good at prices given by $p_i^S = \frac{F}{|S|} + c_i$ for $i \in S$. We iterate this until all agents who are getting the offer accepts it, or all agents refused an offer.

*EFC* is GSP (see below), therefore by Proposition 1 it is not efficient. To illustrate this, consider the utility profile $\tilde{u} = (c_1 + \frac{F}{n} - \varepsilon, c_2 + \frac{F}{n-1} - \varepsilon, c_3 + \frac{F}{n-2} - \varepsilon, \ldots, c_n + F - \varepsilon)$ for $\varepsilon > 0$ small. At $\tilde{u}$ the mechanism does not serve any agent. Indeed, since $\tilde{u}_1 < c_1 + \frac{F}{n}$, then agent 1 declines the offer in the first iteration. Similarly, $\tilde{u}_2 < c_2 + \frac{F}{n-1}$ declines it in the second iteration, etc. However, this is clearly inefficient because by serving all agents we get a surplus $\sum_{i=1}^{n} \frac{F}{i} - F - n\varepsilon \approx \sum_{i=1}^{n} \frac{F}{i} - F > 0$.

Notice that in the above mechanism, prices are cross-monotonic, that is they increase as coalition decreases ($p_i^T \leq p_i^S$ for all $S \subset T$). In general, any set of prices that are cross-monotonic defines a GSP mechanism (see Juarez[2007] for details). For instance, an alternative set of cross-monotonic prices may divide the total cost equally, $p_i^T = \frac{C(T)}{|T|} = \frac{F + \sum_{k \in T} c_k}{|T|}$. However, by Proposition 1, any cross-monotonic mechanism will still be inefficient. We compare this inefficiency by the size of the worst absolute surplus loss (*WAL*), that is by the maximum difference between the efficient surplus and the surplus of the mechanism, where the maximum is taken over all utility profiles (see Moulin and Shenker[2001] and Juarez[2008b] for precise definition). *EFC* has the smallest loss among all GSP mechanisms, including those that do not cover the cost exactly.

**Proposition 2 (Moulin and Shenker(2001), Juarez(2008b)).** *The EFC mechanism is GSP and has a worst absolute surplus loss equal to $\sum_{i=1}^{n} \frac{F}{i} - F$. Moreover it has* *the smallest worst absolute surplus loss among all feasible GSP mechanisms for the single facility location problem.*

## 3 Generalized scheduling problem

In the scheduling problem, every user in $N = \{1, \ldots, n\}$ has a job of the same size. There is a server which can process the jobs, one at a time, at a marginal time (cost) equal to $t_1, \ldots, t_n$. That is the first job is processed in $t_1$ units of time. The second job takes $t_2$ units, but it has to wait until job 1 is finished, so the waiting time to finish the second job is $t_1 + t_2$ units of time, etc.

Notice that the technology to process the jobs may not be linear. Even though the jobs are of the same size, the technology may get more efficient as more jobs are being processed ($t_1 \geq t_2 \geq \cdots \geq t_n$), or more inefficient ($t_1 \leq t_2 \leq \cdots \leq t_n$), or neither.

Agent $i$ has a private monetary valuation $u_i$ for the job to get done. For simplicity, we also assume that all the agents value their time equally. Thus if agent $i$ waits $t_i$ units of time until his job is completed, his net utility is $u_i - t_i$.

A mechanism in this economy elicits from the agents the vector of utilities $(u_1, \ldots, u_n)$, decides what coalition of agents are served, and makes the respective transfers of money between them.

The total cost of serving $k$ jobs is $C(k) = kt_1 + (k-1)t_2 + \cdots + 2t_{k-1} + t_k$.

The *efficient mechanism* is achieved by giving priority to the agents with the highest utility. It collects an efficient surplus equal to $\sum_k (u_k - T_k)_+$ where $u_1 \geq u_2 \geq \cdots \geq u_n$, $T_k = \sum_{j=1}^k t_j$ and $(g)_+ = \max\{g, 0\}$. By fairness reasons, the efficient mechanism is not implemented. As in Proposition 1, there is no mechanisms that is efficient and GSP in this economy.

The *priority mechanism (PRIO)* serves the agents following an arbitrarily priority order. That is, for the priority order $1, \ldots, n$, we offer agent 1 to be processes in $t_1$ units. Agent 2 is offered to be processed in $t_1 + t_2$ units if 1 accepts, or in $t_1$ if 1 does not accept. And similarly for the following agents. *PRIO* is GSP. Indeed, notice agent 1 is offered a unit of good at a price that is independent on the valuation of the other agents. If the true utility of agent 1 is $u_1$, $u_1 > t_1$ he will never misreport a utility less than $t_1$. Similarly, if the true utility of agent 1 is $u_1$, $u_1 < t_1$ then he will never misreport a utility bigger than $t_1$. Thus agent 1 will never profit from misreporting. An easy induction argument shows that there is no group of agents which profits from misreporting.

Only the priority mechanisms are GSP, budget-balanced and meet a consumer-sovereign property (see Moulin[1999]).

The *sequential average cost mechanism (SAC)* offers units of good to the agents sequentially at price equal to average cost. That is, for an arbitrarily order of the agents, $1, \ldots, n$. Agent 1 is offered a unit of good at price $AC(n) = \frac{nt_1 + (n-1)t_2 \cdots + t_n}{n}$. Agent 2 is offered a unit of good at price $AC(n)$ is 1 accepts, or at price $AC(n-1)$

if 1 does not accept, etc. Similarly to *PRIO*, *SAC* is GSP. It minimizes the WAL among all feasible GSP mechanisms.

**Proposition 3 (Juarez(2008b)).** *The SAC mechanism is GSP and has the smallest worst absolute surplus loss among all feasible GSP mechanisms.*

# 4 More general results and related literature

Proposition 1 generalizes to any non-additive cost function. Juarez[2008b] shows that for any non-additive cost function, there is no mechanism that is simultaneously efficient and GSP. If the cost function is additive, a fixed price mechanism, where every agent is offered a unit of good at a price equal to the increment in marginal cost (which is constant by the additivity of the cost function) is GSP and efficient.

The design of GSP cost-sharing mechanisms was first discussed by Moulin [1999]. The paper characterizes cross-monotonic mechanisms similar to Proposition 2 by GSP, budget-balance, voluntary participation, nonnegative transfers and strong consumer sovereignty. Juarez[2007] also characterizes cross-monotonic mechanisms by GSP and a weak continuity condition.

Proposition 2 generalizes to any submodular cost function. Moulin and Shenker [2001] evaluate the trade-offs between efficiency and budget-balance. The Shapley value cross-monotonic mechanism (the analog to the *EFC* mechanism for submodular cost functions) is characterized there by GSP, budget-balance, voluntary participation, nonnegative transfers, strong consumer sovereignty and minimizing the WAL. Juarez[2008b] shows that the same result generalizes even without assuming budget-balance. Roughgarden et al.[2008b] shows that this mechanism is also optimal among the class of strategyproof and 'weakly-monotonic' mechanisms using the worst 'relative' gain measure.

Roughgarden et al.[2006a, 2006b], Pa'l et al.[2003] and Immorlica et al.[2005] consider cross-monotonic mechanisms when the cost function is not submodular. Roughgarden et al.[2006a] uses submodular cross-monotonic mechanisms to approximate budget-balance when the actual cost function is not submodular. Immorlica et al.[2005] shows that new cross-monotonic mechanisms emerge when consumer sovereignty is relaxed.

Proposition 3 generalizes to any cost function that is supermodular. Juarez[2008b] shows that SAC cuts the efficiency loss by half with respect to the optimal budget balanced mechanism for any cost function that is symmetric and supermodular. Besides this result, very little has been said in the literature on mechanisms that generate cost functions that are not submodular. Sequential mechanisms similar to *SAC* are discussed by Moulin[1999] who imposes budget-balance for a supermodular cost function, and by Juarez[2007] who characterizes sequential mechanism by GSP and a weak continuity condition.

Roughgarden et al.[2008a] uncovers a very clever class of weakly GSP mechanisms that are neither cross-monotonic nor sequential. This class contains sequential and cross-monotonic mechanisms, as well as hybrid mechanisms. They apply these mechanisms to the vertex cover and Steiner tree cost sharing problems to improve the efficiency of algorithms derived from cross-monotonic mechanisms. Juarez[2008c] develops a model where indifferences are ruled out. For instance, agents report an irrational number and payments are rational. It turns out that the class of GSP mechanisms becomes very large. In particular, it contains mechanisms very different to cross-monotonic and sequential mechanisms (and also those discussed by Roughgarden et al.[2008a]). Juarez[2008c] provides three equivalent characterizations of the GSP mechanism in this economy, two of which are generalizations of the cross-monotonic and sequential mechanisms discussed in this paper.

# 5 Open problems

The problem of designing mechanism in the prior-free context is far from being solved. The study of worst-case scenarios and robust mechanism design is very promising not only in the network economic problems mentioned above, but also in other contexts like auctions, cost-sharing, decision theory and social choice theory.

The main limitation of propositions 2 and 3 is that it restrict to cost functions that are submodular and supermodular. For instance, just by adding a second facility in the problem discussed in section 2 generate a cost function that is neither submodular nor supermodular. More interesting and challenging computer science applications like vertex cover problem, set cover problem, metric and non-metric uniform facility location problems generate even more complicated cost functions. Finding optimal mechanisms for the cost functions generated by these problems is an interesting but difficult question.

In some contexts, even worst case measures do not work well. For instance, for the case of prior-free actions, where the objective of the seller is to maximize his profit (instead of the consumer's surplus), traditional worst-case techniques are not informative. Indeed, assume there is one seller with a single good and a single buyer in the economy with private valuation for this good. Now, consider for instance the worst absolute profit-loss measure, that is the difference between the monopolist's profit and the profit of the mechanism at a given profile, and compute the supremum of this difference for all utility profiles. This measure is not informative. To see this, first notice that any strategyproof mechanism should by a posted price mechanism at some given price $p$. If the valuation of the buyer is $b$, $b > p$, then the regret of the seller by charging $p$ is $b - p$. As $b$ grows, so does the regret, thus the worst absolute profit-loss is infinity. Hence, what are good measures of the seller's profit in the prior-free setting and what are the optimal auctions using this measure? Work by Baliga and Vohra[2003], Goldberg, Hartline, Karlin, Saks and Wright[2005] and Bergemann and Schlag[2007, 2008] are starting points.

Finally, GPS is a robust property that rules out coordination in the bids of the agents, but does not rule out coordination in the bids and monetary transfers between them (a very strong form of collusion). Several types of collusion-proof mechanisms are discussed in Juarez[2008c]. The designing of markets and other institutions in a robust manner is an interesting line of research that is far from being exhausted.

# References

1. Baliga S. and Vohra R. (2003) Market Research and Market Design. The B.E. Journal in Theoretical Economics 3.
2. Bergemann D., Schlag K. (2007) Robust Monopoly Pricing. Discussion Paper 1527R, Cowles Foundation for Research in Economics, Yale University.
3. Bergemann D., Schlag K. (2008) Pricing without Priors, Journal of the European Economic Association (April-May 2008), 6(2-3): 560-569.
4. Goldberg A., Hartline J., Karlin A., Saks M., Wright A. (2005) Competitive Auctions. Mimeo Mcrosoft
5. Immorlica N., Mahdian M., Mirrokni W. (2005) Limitations of cross-monotonic cost sharing schemes, Symposium on Discrete Algorithms (SODA).
6. Juarez R. (2007) Group strategyproof cost sharing: The role of indifferences. Mimeo University of Hawaii
7. Juarez R. (2008a) The worst absolute surplus loss in the problem of commons: Random Priority vs. Average Cost. Economic Theory 34.
8. Juarez R. (2008b) Optimal group strategyproof cost sharing: budget balance vs. efficiency. Mimeo University of Hawaii
9. Juarez R. (2008c) Collusion-proof cost sharing. Mimeo University of Hawaii
10. Moulin H. (1999) Incremental Cost Sharing: characterization by coalitional strategy-proofness. Social Choice and Welfare **16**.
11. Moulin H. (2008) The price of anarchy of serial cost sharing and other methods. Econ. Theory 36.
12. Moulin H., Shenker S. (2001) Strategyproof sharing of submodular costs: budget balance versus efficiency. Economic Theory **18**.
13. Myerson R. (1981) Optimal Auction Design. Mathematics of Operation Research 6.
14. M. P'al, E. Tardos (2003) Group strategyproof mechanisms via primal-dual algorithms, In Proceedings of 44th Annual IEEE Symposium on Foundations of Computer Science (FOCS).
15. Roughgarden T. Sundararajan M. (2006a) Approximately efficient cost-sharing mechanisms. Mimeo Stanford
16. Roughgarden T., Sundararajan M. (2006b) New TradeOffs in Cost Sharing Mechanisms, STOC.
17. T. Roughgarden, Mehta A., Sundararajan M. (2008a) Beyond Moulin Mechanisms, Forthcoming Games and Economic Behavior.
18. Roughgarden T., Dobzinski S., Mehta A., and Sundararajan M.(2008b) Is Shapley Cost Sharing Optimal? Symposium on Algorithmic Game Theory.

# Monitoring Web Resources Discovery by Reusing Classification Knowledge

Byeong Ho Kang and Yang Sok Kim

{bhkang, yangsokk}@utas.edu.au, University of Tasmania, Tasmania, Australia

**Abstract** Any automated client pull systems, such as web monitoring systems, Web services, or RSS systems, require resources that publish relevant information. These resources may be discovered manually, but this is not ideal. Public search engines may be used to find these resources by submitting appropriate queries. In the previous research, we proposed a search query formulation method that reuses MCRDR (Multiple Classification Ripple-Down Rules) classification knowledge bases. However, all search results may not relevant web resources, because they may not resources, but web pages linked from the candidate resources. Therefore, it is necessary to develop methods that locate candidate monitoring web resources from search results. This paper summarizes heuristics that were obtained user study, which will be used in this automated resource location process.

## 1 Introduction

Nowadays information is published on the Web and people wish to get relevant information promptly. Various automated client pull systems, such as web monitoring systems, Web services, or RSS (Really Simple Syndication) systems, were proposed to support this kind of information need. Finding and registering relevant resources, such as monitoring Web pages and RSS feed pages, is a necessary process in any automated client pull systems. It may be conducted manually; however, this is not an ideal solution. Web search engines may be used to locate monitoring resources for this purpose, as they provide significantly large volumes of web pages collected from various web resources. To use search engines for this purpose, the following processes should be conducted: Firstly, appropriate search queries should be submitted to the search engines. We investigated search query formulation method that reuses the MCRDR (Multiple Classification Ripple–Down Rules) document classification knowledge bases. Among various types of search queries, the 'rule-condition-based query' was proved as the best query generation method [1]. Secondly, once relevant search results are obtained from search engines, it is necessary to locate monitoring web pages using them, because most search results are not *monitoring web pages*, but *information pages* published from their monitoring web pages. Lastly, it is necessary to determine whether or not the identified monitoring web pages are to be used for collecting in-

H. Liu et al. (eds.), *Social Computing and Behavioral Modeling*,
DOI: 10.1007/978-1-4419-0056-2_17, © Springer Science + Business Media, LLC 2009

formation by the automated client pull systems, as not all the monitoring resources obtained from the search results publish relevant information. In these processes, this paper focuses on the last two processes. Firstly, this paper discusses how the users locate their monitoring web pages using search results. For this purpose, a simulated monitoring web page location experiment was conducted to evaluate monitoring web page location heuristics. This experiment was extended from [2]. For the given search results, each participant was requested to classify them into the specific web page type and to nominate candidate monitoring web pages' URLs. Secondly, this paper discusses how the users determine the usefulness of the candidate monitoring web pages. Finally, this paper proposes an automated monitoring web page recommendation method that reuses the MCRDR classification knowledge base. This paper consists of the following contents: Section 2 summarizes background information and related research. Section 3 explains the experimental methodology. Section 4 summarizes simulated monitoring web page locating results. Conclusions are provided in Section 5.

## 2 Related Studies

Web page type classification endeavours to classify web pages into one of several specific page types. This differs from web document classification, which aims to classify documents into predefined category using the similarity between documents' contents or their hyperlink structures [3, 4, 5]. Web page type classification formerly gained little attention from researchers, when it is compared with web document classification. Matsuda and Fukushima [6] identified web page type classification problems, and tried to solve them by describing structural characteristics. Glover et al. [7] manually classified a large number of documents to collect negative training data into "personal homepage" and "call-for-paper" categories using SVM (Support Vector Machines) with fixed sets of features. Elsas and Efron [8] proposed HTML tag based metrics for web page type classification. They used the table tag ratio, anchor text ratio, and text per table data tag to develop thresholds that classify web pages into (data) table, index/table of contents, and content pages. Web page type classification usually depends on the particular tasks. For example, Glover et al. [7] stated that "A personal homepage is a difficult concept to define objectively. The definition we used is a page made by an individual (or family) in an individual role, with the intent of being the entry point for information about the person (or family)." (p.6). Therefore, the web page classification method should take into account tasks specific to this research. As our research focuses on finding candidate web monitoring pages, web page type classification needs to be conducted with this purpose in mind and accordingly, the classification approach should be developed in relation to this task.

Backlinks of information pages, provided by search engines, might be considered as monitoring web pages. Commercial search engines provide search options for finding backlinks of a specific web page. However, there are some limitations in using backlinks as a method for finding monitoring pages. Firstly, the monitoring page is not necessarily the backlink page, as the latter may contain the parent page, but may also include many other web pages as backlink pages. Secondly, search engines do not provide all backlinks, the provision of backlinks depends totally on the search engine companies' decision [9]. Therefore, it is necessary to research the heuristics for finding the monitoring page from a given search result page. However, the previous web monitoring research did not suggest any systematic approaches for locating monitoring web pages, rather registration was conducted manually [10, 11, 12, 13].

## 3 Research Design

The systems and the processes used for the monitoring web page locating experiment are illustrated in the left side of Figure 1. The Web page Finding System (WFS) creates search queries from the MCRDR document classification knowledge and retrieves web pages from the search engine's database by submitting these queries. By using the Web page Evaluation System (WES) participants were required to evaluate the collected web pages, to classify their type, to find the candidate monitoring page of the document page, and to recommend the final monitoring page.

In total, 40 Masters and Honours students from the School of Computing at the University of Tasmania participated in this experiment. Each participant was required to locate the monitoring web pages from 210 web pages that were randomly sampled from the rule-condition-based search results. In total, 8,400 web pages (210 web pages $\times$40 participants) were utilised. The user interface of the WES is illustrated in the right side of Figure 1. The participant was required to select the web page type; 'index page' (monitoring web page), 'document page', or 'ad-hoc page'. If the answer to the first question was 'index', the system interrogated the user by asking whether or not he/she wanted to recommend this page as a monitoring page. If the participant chose 'ad-hoc page' in the first question, the system saved the page type and finished the evaluation process. Finally, if the participant chose the 'document page' as the answer to the first question, the system asked the participant whether or not he/she found the parent page (the candidate monitoring page) of the document page. If the participant found the parent page, he/she was required to fill the URL into the 'Fill in parent page URL' field and answer how he/she found the page, and whether or not the page should be monitored. Otherwise, he/she selected the negative answer ("No. I cannot find the monitoring page (parent page).").

In addition, the candidate monitoring web pages that each participant chooses to be monitored in the above evaluation were automatically evaluated by classification knowledge bases, which were obtained in the previous document classification research [1]. The MCRDR Document classification knowledge base can be used to decide whether or not a web page is related to the user's interest, assuming that a candidate monitoring page should be monitored if it is fired by a knowledge base. The candidate monitoring web pages (parent pages) that obtained from the above experiment were processed by ten MCRDR classification knowledge bases used for [1] were reused for this purpose. These recommendations were compared with the above evaluation results.

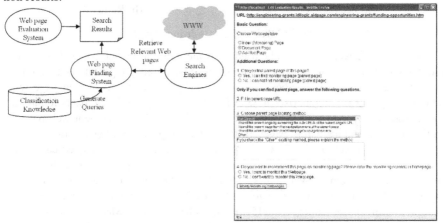

*Figure 1 Experimental Process and the Systems Used*

# 4 Experimental Results

The web page classification results of participants are illustrated in Figure 2, where (a) shows classification frequency by each web page type and (b) displays its relative ratio. About half of all search results (52%) are document pages and 30% are monitoring pages, while 18% are ad-hoc pages. This means most of the search results (82%) were clearly classified by the users and this percentage may be the maximum level that an automated web page classification system can attain. In addition, about a third of the search results (30%) are directly classified as candidate monitoring pages, while about half of the search results (52%) needed further processing to find the candidate monitoring web pages.

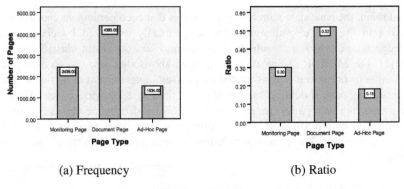

(a) Frequency                          (b) Ratio

*Figure 2 Web Page Classification Results*

Document type web pages can be used to locate monitoring web pages. Figure 3 illustrates candidate monitoring page finding results, based on parent page finding, where Figure 3 (a) shows whether or not a parent page of the given web page was found and Figure 3 (b) summarises how to find the parent page. Participants found the parent pages of a total of 3,518 web pages (80% of the 4,380 document web pages), but they did not find the parent pages of 862 document pages (see Figure 3 (a)). Therefore, a total 6,004 web pages (71% of 8400 web pages) have an associated candidate monitoring web page. Figure 3 (b) summarises how the participants located the parent pages of 3,518 document web pages. They found 44% parent pages by using the 'navigation links in the current page', followed by the 'sub-URL' method (36%), 'navigation links in the homepage' (15%), and 'others' (4.0%). Major methods classified as 'others' were 'breadcrumbs' and 'search'. Though the 'Sub-URL' method was suggested as a preferable strategy to the 'link in the current page' method in the previous user study results, the experimental classification shows that the latter approach was used more frequently than the former.

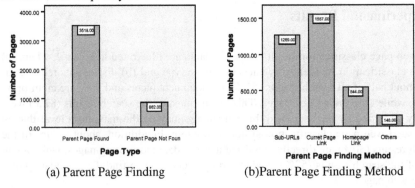

(a) Parent Page Finding          (b)Parent Page Finding Method

*Figure 3 Candidate Monitoring Web Page Finding*

Not all candidate monitoring web pages are worth monitoring. An additional question was given to the participants asking whether or not they wanted to use the candidate monitoring web page for monitoring purposes under the given web monitoring scenario. Figure 4 summarises the monitoring recommendations of the candidate monitoring web pages. On average, 58% of the candidate monitoring pages was recommended for monitoring by the participants, as is illustrated in Figure 4 (a). The distribution of recommendation ratios is displayed in Figure 4 (b), where the horizontal axis represents the proportion of the candidate monitoring web pages that were recommended as final monitoring web pages. The ratios significantly vary between participants as illustrated in Figure 4 (b).

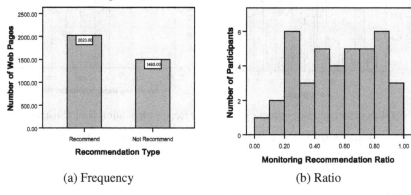

(a) Frequency                                      (b) Ratio

*Figure 4 Monitoring Web Page Recommendations by Participants*

Overall, 58% of the candidate monitoring web pages were recommended as final monitoring web pages. As the generic top-level domains (gTLD) such as ".com", ".org", and ".gov" were widely used, the proportion of pages from these domains was very high. One salient feature of these results was that the ratios of the number of recommended web pages (B) to the number of all parent pages found (A) were significantly higher in Australian ccTLD (Country-Code Top-Level Domains) such as "gov.au"(81%), "com.au"(66%), "org.au"(78%), "net.au"(87%), and "asn.au" (80%) than in other domains. This result was affected by the web monitoring scenario, given that the experiment focused on Australian Government relevant information. The number of recommended web pages became less as the page depth increased. For example, 41% of all the recommended web pages were zero (0) page depth – that is, at the host URL. However, the ratio of recommended pages to the number of the pages found (B/A) increased as the page depth increased, because the high depth web pages usually contained specific contents compared to the low depth web pages.

Figure 5 illustrates monitoring web page recommendation results from ten knowledge bases, where Figure 5 (a) illustrates each knowledge base's recommendation frequency and Figure 5 (b) shows how the monitoring ratios were distributed. The average recommendation was 1,976 web pages, which was 56% of all search results that had a candidate monitoring page (parent page). Even though the overall ratio of knowledge based recommendation was similar to the experimental participants results, its distribution was significantly different as illustrated in Figure 5 (b). Whereas the latter shows a long tailed distribution, the former clustered between 40% and 90%.

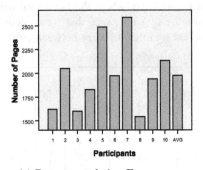 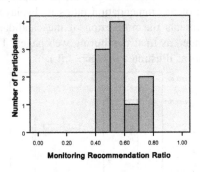

(a) Recommendation Frequency      (b) Recommendation Ratio Distribution

*Figure 5 Monitoring Web Page Recommendations with Knowledge Base*

# 5 Conclusions

This research investigated heuristics for locating monitoring web pages using the search results obtained from commercial search engines, with the queries generated by reusing MCRDR document classification knowledge. The web page classification experiment illustrated that 50% of search results were document pages, 28% monitoring pages, and 14% ad-hoc pages. Therefore, most search results (about 80%) can be used to locate monitoring web pages. Candidate monitoring web pages were found from 80% of document pages. The 'links in the current page' strategy was used most frequently to find the candidate monitoring web pages, followed by the 'Sub-URL' strategy and the 'links in the home page' strategy. Among the candidate monitoring web pages, about 55% were recommended as monitoring pages. Update regularity, relevance to the user's interest, and the number of links on the page were suggested as the most significant factors determining which pages should be monitored. Human heuristics were identified for locating monitoring web pages in this research. Further research will build on these results to implement an automated monitoring web page locating system.

**Acknowledgement** This work was supported by the Asian Office of Aerospace Research and Development (AOARD).

# References

[1]   Kim, Y.S. and B.H. Kang. *Search Query Generation with MCRDR Document Classification Knowledge.* in *EKAW 2008 - 16th International Conference on Knowledge Engineering and Knowledge Management Knowledge Patterns.* 2008. Acitrezza, Catania, Italy.

[2]   Kim, Y.S. and B.H. Kang. *A Study on Monitoring Web Page Locating Heuristics.* in *The 2008 International Conference on Information and Knowledge Engineering (IKE'08 ).* 2008. Monte Carlo Resort, Las Vegas, Nevada, USA.

[3]   Sebastiani, F., *Machine learning in automated text categorization.* ACM Computing Surveys, 2002. **34**(1): p. 1-47.

[4]   Sebastiani, F., *Text categorization,* in *The Encyclopedia of Database Technologies and Applications,* L.C. Rivero, J.H. Doorn, and V.E. Ferraggine, Editors. 2005, Idea Group Publishing: Hershey, US.

[5]   Sebastiani, F., *Text categorization,* in *Text Mining and its Applications,* A. Zanasi, Editor. 2004, WIT Press, Southampton, UK. p. pp. 109--129.

[6]   Matsuda, K. and T. Fukushima. *Task-oriented world wide web retrieval by document type classification.* in *the eighth international conference on Information and knowledge management.* 1999. Kansas City, Missouri, United States: ACM   New York, NY, USA.

[7]   Glover, E.J., G.W. Flake, S. Lawrence, W.P. Birmingham, A. Kruger, C.L. Giles, and D.M. Pennock. *Improving Category Specific Web Search by Learning Query Modifications.* in *SAINT 2001.* 2001. San Diego, California: IEEE Computer Society.

[8]   Elsas, J. and M. Efron. *HTML tag based metrics for use in web page type classification.* in *American Society for Information Science and Technology Annual Meeting.* 2004. Providence, Rhode Island, USA.

[9]   Wilson, R.F., *Google's Index Shows Only a Few Backlinks.* 2006.

[10]  Boyapati, V., K. Chevrier, A. Finkel, N. Glance, T. Pierce, R. Stockton, and C. Whitmer. *ChangeDetector[tm]: a site-level monitoring tool for the WWW.* in *Eleventh International World Wide Web Conference (WWW 2002).* 2002. Hawaii, USA.

[11]  Douglis, F. and T. Ball. *Tracking and Viewing Changes on the Web.* in *USENIX Annual Technical Conference.* 1996.

[12]  Liu, L., W. Tang, D. Buttler, and C. Pu, *Information Monitoring on the Web:A Scalable Solution.* World Wide Web Journal, 2002. **5**(4): p. 263-304.

[13]  Pandey, S., K. Dhamdhere, and C. Olston. *WIC: A General-Purpose Algorithm for Monitoring Web Information Sources.* in *30th VLDB Conference.* 2004. Toronto, Canada.

# Finding Influential Nodes in a Social Network from Information Diffusion Data

Masahiro Kimura[1], Kazumi Saito[2], Ryohei Nakano[3], and Hiroshi Motoda[4]

[1]kimura@rins.ryukoku.ac.jp, Ryukoku University, Shiga, Japan
[2]k-saito@u-shizuoka-ken.ac.jp, University of Shizuoka, Shizuoka, Japan
[3]nakano@cs.chubu.ac.jp, Chubu University, Aichi, Japan
[4]motoda@ar.sanken.osaka-u.ac.jp, Osaka Univesity, Osaka, Japan

**Abstract** We address the problem of ranking influential nodes in complex social networks by estimating diffusion probabilities from observed information diffusion data using the popular independent cascade (IC) model. For this purpose we formulate the likelihood for information diffusion data which is a set of time sequence data of active nodes and propose an iterative method to search for the probabilities that maximizes this likelihood. We apply this to two real world social networks in the simplest setting where the probability is uniform for all the links, and show that the accuracy of the probability is outstandingly good, and further show that the proposed method can predict the high ranked influential nodes much more accurately than the well studied conventional four heuristic methods.

## 1 Introduction

Innovation, hot topics and even malicious rumors can propagate through social networks among people in the form of so-called "word-of-mouth" communications. The rise of the Internet and the World Wide Web accelerates the creation of various large-scale social networks. Therefore, considerable attention has recently been devoted to social networks as an important medium for the spread of information.

Previous work addressed the problem of tracking the propagation patterns of topics or influence through blogspace [1, 5, 10], and studied strategies for removing nodes to prevent the spread of some undesirable information through a network, for example, the spread of a computer virus through an email network [2, 11]. A widely-used fundamental probabilistic model of information diffusion through a network is the *independent cascade (IC) model* [6, 5]. Using this model, the problem of finding a limited number of nodes that are effective for the spread of information [6, 8] has been extensively investigated. This combinatorial optimization problem is called the influence maximization problem. This problem was also investigated in a different setting (a descriptive probabilistic model of interaction) [4, 13]. Further, yet another problem of minimizing the spread of undesirable information by blocking a limited number of links in a network [9] has recently been addressed. In this paper, we also explore information diffusion phenomena for the IC model in a given network.

H. Liu et al. (eds.), *Social Computing and Behavioral Modeling*,
DOI: 10.1007/978-1-4419-0056-2_18, © Springer Science + Business Media, LLC 2009

Overall, finding influential nodes in a social network is one of the most central problems in the field of social network analysis. There exist several methods for ranking nodes on the basis of the network structure [15]. We also address this problem, but from a different angle. We propose a method for extracting influential nodes by ranking nodes in terms of *influence degrees* for the IC model on the basis of the observed data of information diffusion in the network. The IC model is equipped with parameters. More specifically, the *diffusion probability* must be specified for each link in the network in advance. We estimate the probabilities so that the likelihood of obtaining the observed set of information diffusion data is maximized by an iterative algorithm (EM algorithm). Using two real world networks: the blog and Wikipedia networks, we first evaluate the accuracy of the diffusion probabilities and then use the estimated model to find the influential nodes and compare the results with the ground truth as well as the results that are obtained by using four strategies, each with a different heuristic, showing that the proposed method far outperforms the conventional methods.

The rest of the paper is organized as follows. The proposed method is formulated as a machine learning problem in Sect. 2, and the experimental results together with the experimental settings are given in Sect. 3, followed by some discussion of how the probabilities affect the influential nodes in Sect. 4. We conclude this paper by summarizing our findings in Sect. 5.

## 2 Proposed Method

### 2.1 Problem Formulation and Extraction Method

For a given directed network (or equivalently graph) $G = (V, E)$, let $V$ be a set of nodes (or vertices) and $E$ a set of links (or edges), where we denote each link by $e = (v, w) \in E$ and $v \neq w$, meaning there exists a directed link from a node $v$ to a node $w$. For each node $v$ in the network $G$, we denote by $F(v)$ a set of child nodes of $v$ as follows: $F(v) = \{w; (v, w) \in E\}$. Similarly, we denote by $B(v)$ a set of parent nodes of $v$ as follows: $B(v) = \{u; (u, v) \in E\}$.

In the IC model, for each directed link $e = (v, w)$, we specify a real value $p_{v,w}$ with $0 < p_{v,w} < 1$ in advance. Here $p_{v,w}$ is referred to as the *diffusion probability* of link $(v, w)$. The diffusion process proceeds from a given initial active set $D(0)$ in the following way. When a node $v$ first becomes active at time-step $t$, it is given a single chance to activate each currently inactive child node $w$, and succeeds with probability $p_{v,w}$. If $v$ succeeds, then $w$ will become active at time-step $t + 1$. If multiple parent nodes of $w$ first become active at time-step $t$, then their activation attempts are sequenced in an arbitrary order, but all performed at time-step $t$. Whether or not $v$ succeeds, it cannot make any further attempts to activate $w$ in subsequent rounds. The process terminates if no more activations are possible.

For a given set of diffusion probabilities, $\Theta = \{p_{v,w}; (v,w) \in E\}$, and an initial active node $v$, we define the *influence degree* , denoted by $\sigma(v; \Theta)$, as the expected number of active nodes. Our problem of finding influential nodes is formulated as a node ranking problem based on the influence degree $\sigma(v; \Theta_0)$, where $\Theta_0$ means a set of the true diffusion probabilities. In practice settings, however, the true diffusion probability set $\Theta_0$ is not available. Thus, we consider to utilize their probabilities $\widehat{\Theta}$ estimated from past information diffusion histories observed as sets of active nodes. Then we need to evaluate the ranking similarity between two sorted node lists according to $\sigma(v; \Theta_0)$ and $\sigma(v; \widehat{\Theta})$.

## 2.2 Probability Estimation Method

Let $D = D(0) \cup D(1) \cup \cdots \cup D(T)$ be an information diffusion result, where $D(t)$ is the set of nodes that have become active at time $t$. When $v \in D(t)$ and $w \in D(t+1) \cap F(v)$ hold for some link $e = (v,w)$, it is possible that the node $v$ succeeded in activating the node $w$ via the link $e$. However, since we should consider possibilities that some other nodes $v' \in D(t) \cap B(w)$ also succeeded in activating the node $w$, we need to calculate the probability that the node $w$ becomes active at time $t+1$ as follows: $P(w; t+1) = 1 - \prod_{v \in B(w) \cap D(t)} (1 - p_{v,w})$. Here note that if $w \in D(t+1)$, it is guaranteed that $D(t) \cap B(w) \neq \emptyset$.

We set $C(t) = D(0) \cup \cdots \cup D(t)$. Note that $C(t)$ is the set of active nodes at time $t$. When $v \in D(t)$ and $w \in F(v) \setminus C(t+1)$ hold, we know that the node $v$ definitely failed to activate the node $w$ via the link $e$. Clearly, when $v \in D(t)$ and $w \in F(v) \cap C(t)$ hold, as well as $v \notin D$, no information is available about the trial with respect to the link $e = (v,w)$. Therefore, we can define the likelihood function with respect to $\Theta = \{p_{v,w}\}$ as follows:

$$\mathscr{L}(\Theta; D) = \prod_{t=0}^{T-1} \prod_{w \in D(t+1)} \left(1 - \prod_{v \in B(w) \cap D(t)} (1 - p_{v,w})\right) \prod_{t=0}^{T} \prod_{v \in D(t)} \prod_{w \in F(v) \setminus C(t+1)} (1 - p_{v,w}).$$

Let $\{D_m; 1 \leq m \leq M\}$ be an observed data set of $M$ independent information diffusion results. Then we can define the following objective function with respect to $\Theta$:

$$\mathscr{J}(\Theta) = \sum_{m=1}^{M} \log \mathscr{L}(\Theta; D_m). \tag{1}$$

Thus, our problem is to obtain the set of information diffusion probabilities $\Theta$, which maximizes Equation (1). For this estimation problem, we have already proposed an estimation method based on the Expectation-Maximization algorithm in order to stably obtain its solutions [14].

In order to evaluate fundamental abilities of our method, in this paper, we consider the simplest case that all links have the same diffusion probability $p$. Note that this problem setting has been widely adopted in many previous experiments

[6, 8, 9], and the formulation is valid for more general cases in which there is no such restriction.

# 3 Experiments

## 3.1 Experimental Settings

We employed two sets of large real networks used in [9], the blog and Wikipedia networks, which exhibit many of the key features of social networks. These are bidirectional networks. The blog network had 12,047 nodes and 79,920 directed links, and the Wikipedia network had 9,481 nodes and 245,044 directed links. As stated before, in our preliminary experiments, we assumed the simplest case where the diffusion probability is uniform throughout the network, and set the value $p$ as follows: $p = 0.1$ for the blog network and $p = 0.01$ for the Wikipedia network. We evaluated the influence degrees $\{\sigma(v); v \in V\}$ using the method of [8] with the parameter value $10,000$, where the parameter represents the number of bond percolation processes (we do not describe the method here due to the page limit). The average value and the standard deviation of the influence degrees was 87.5 and 131 for the blog network, and 8.14 and 18.4 for the Wikipedia network.

In the learning stage, a training sample was an information diffusion path $D = D(0) \cup D(1) \cup \cdots \cup D(T)$ which is a sequence of the active nodes starting from a randomly selected initial active node. We used $M$ training samples for learning the propagation probability, where $M$ is a parameter.

## 3.2 Comparison Methods

We compared the proposed method with four heuristics from social network analysis with respect to the predictive capability of high ranked influential nodes.

First, "degree centrality", "closeness centrality", and "betweenness centrality" are commonly used as influence measure in sociology [15], where the degree of node $v$ is defined as the number of links attached to $v$, the closeness of node $v$ is defined as the reciprocal of the average distance between $v$ and other nodes in the network, and the betweenness of node $v$ is defined as the total number of shortest paths between pairs of nodes that pass through $v$.

We also consider measuring the influence of each node by its "authoritativeness" obtained by the "PageRank" method [3], since this is a well known method for identifying authoritative or influential pages in a hyperlink network of web pages. This method has a parameter $\varepsilon$; when we view it as a model of a random web surfer, $\varepsilon$ corresponds to the probability with which a surfer jumps to a page picked uniformly at random [12]. In our experiments, we used a typical setting of $\varepsilon = 0.15$.

## 3.3 Experimental Results

First, we examined the learning performance of propagation probability by the proposed method. Let $p_0$ be the true value of propagation probability, and let $\hat{p}$ be the value of propagation probability estimated by the proposed method. We evaluated the learning performance in terms of the error rate $\mathscr{E} = |p_0 - \hat{p}|/p_0$.

**Table 1** Learing performance of propagation probability.

| Results for the blog network | | Results for the Wikipedia network | |
| --- | --- | --- | --- |
| $M$ | $\mathscr{E}$ | $M$ | $\mathscr{E}$ |
| 20 | 0.036 (0.024) | 20 | 0.138 (0.081) |
| 40 | 0.018 (0.014) | 40 | 0.109 (0.066) |
| 60 | 0.016 (0.007) | 60 | 0.080 (0.041) |
| 80 | 0.009 (0.006) | 80 | 0.047 (0.018) |
| 100 | 0.006 (0.004) | 100 | 0.021 (0.013) |

Table 1 shows the average value of $\mathscr{E}$ and the standard deviation in parenthesis for the number of training samples, $M$, where we performed the same experiment five times independently. Our algorithm can converge to the true value efficiently when there is a reasonable amount of training data. The results are better for a larger value of diffusion probability. The results demonstrate the effectiveness of the proposed method.

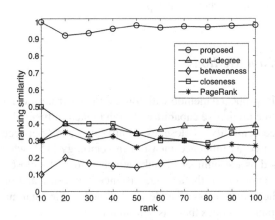

**Fig. 1** Performance comparison in extracting influential nodes for the blog network.

Next, in terms of ranking for extracting influential nodes from the network $G = (V,E)$, we compared the proposed method with the out-degree, the betweenness, the closeness, and the PageRank methods. For any positive integer $r$ $(\leq |V|)$,

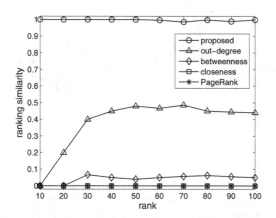

**Fig. 2** Performance comparison in extracting influential nodes for the Wikipedia network.

let $L_0(r)$ be the true set of top $r$ nodes, and let $L(r)$ be the set of top $r$ nodes for a given ranking method. We evaluated the performance of the ranking method by the *ranking similarity* $F(r)$ at rank $r$, where $F(r)$ is defined by $F(r) = |L_0(r) \cap L(r)|/r$. We focused on ranking similarities at high ranks since we are interested in extracting influential nodes. Figures 1 and 2 show the results for the blog and the Wikipedia networks, respectively. Here, circles, triangles, diamonds, squares, and asterisks indicate ranking similarity $F(r)$ as a function of rank $r$ for the proposed, the out-degree, the betweenness, the closeness, and the PageRank methods, respectively. For the proposed method, we plotted the average value of $F(r)$ at $r$ for five experimental results in the case of $M = 100$. The proposed method gives far better results than the other heuristic based methods for the both networks, demonstrating the effectiveness of the proposed method.

## 4 Discussion

We consider that our proposed ranking method presents a novel concept of centrality based on the information diffusion model, i.e., *the IC model*. Actually, Figs. 1 and 2 show that nodes identified as higher ranked by our method are substantially different from those by each of the conventional methods. This means that our method enables a new type of social network analysis if past information diffusion data are available. Of course, it is beyond controversy that each conventional method has its own merit and usage, and our method is an addition to them which has a different merit in terms of information diffusion.

Here, we do some simple analysis of explaining why it is important to know the diffusion probability in finding the influential nodes. If the probability does not affect the ranking, we don't care about its absolute value. However, a simple anal-

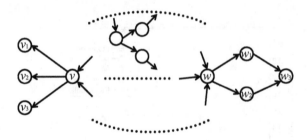

**Fig. 3** An example of network.

ysis reveals that it does affect the node ranking. Note that $\sigma(v;p)$ is a monotonically increasing non negative function of $p$ if $v$'s out degree is non zero. Assume that there are two such nodes $v$ and $w$ that have the following graph structures: $(v,v_1),(v,v_2),(v,v_3) \in E$ and $(w,w_1),(w,w_2),(w_1,w_3),(w_2,w_3) \in E$ (see Fig. 3). The maximum influential degree is 3 for the both nodes $v$ and $w$. The expected values are easily calculated [7] as $\sigma(v;p) = 3p$, and $\sigma(w;p) = 2p + (1 - (1 - p^2)^2) = 2p + 2p^2 - p^4$. Thus, $\sigma(v;p) - \sigma(w;p) = p(1-p)(1-p-p^2)$. From this, if $p < (-1+\sqrt{5})/2, \sigma(v;p) > \sigma(w;p)$. Otherwise, $\sigma(v;p) \leq \sigma(w;p)$. Intuitively, as $p$ gets larger, the influential probability of the nodes reachable in two steps from the starting node becomes larger than that of the nodes reachable in one step, and thus, $w$ that has child nodes in two steps downward has a larger influential degree. Since in general there are many subnetworks like these within a network, it is important to estimate the diffusion probabilities as accurately as possible. We believe that the methods proposed in this paper would contribute to various types of social network analyses.

We note that the analysis we showed in this paper is the simplest case where $p$ takes a single value for all the links in $E$. However, the method is very general. In a more realistic setting we can divide $E$ into subsets $E_1, E_2, ..., E_N$ and assign a different value $p_n$ for all the links in each $E_n$. For example, we may divide the nodes into two groups: those that strongly influence others and those not, or we may divide the nodes into another two groups: those that are easily influenced by others and those not. We can further divide the nodes into multiple groups. If there is some background knowledge about the node grouping, our method can make the best use of it, one of the characteristics of the artificial intelligence approach. Obtaining such background knowledge is also an important research topic in the knowledge discovery from social networks.

# 5 Conclusion

We addressed the problem of ranking influential nodes in complex social networks, given the network topology and the observation data of information diffusion. We formulated how to estimate the diffusion probability of each link from the past information diffusion histories observed as sets of active nodes using the popular information diffusion model, *IC model* as a likelihood maximization problem and derived an efficient iterative EM method to solve it. The results we obtained by applying to two real world networks in the simplest setting where the probability is uniform throughout each network show that 1) the method can estimate the probability accurately when there is enough number of observation sequence data that can be used for training and 2) the ranking of influential nodes predicted by the method far outperforms the other well known heuristic based methods (degree centrality, closeness centrality, betweenness centrality, and authoritativeness).

**Acknowledgements** This work was partly supported by Asian Office of Aerospace Research and Development, Air Force Office of Scientific Research, U.S. Air Force Research Laboratory under Grant No. AOARD-08-4027, and JSPS Grant-in-Aid for Scientific Research (C) (No. 20500147).

# References

1. Adar E, Adamic L (2005) Tracking information epidemics in blogspace. In: WI'05 207–214
2. Albert R, Jeong H, Barabási A L (2000) Error and attack tolerance of complex networks. Nature 406:378–382
3. Brin S, Page L (1998) The anatomy of a large-scale hypertextual web search engine. In: WWW'98 107–117
4. Domingos P, Richardson M (2001) Mining the network value of customers. In: KDD'01 57–66
5. Gruhl D, Guha R, Liben-Nowell D, Tomkins A (2004) Information diffusion through blogspace. In: WWW'04 107–117
6. Kempe D, Kleinberg J, Tardos E (2003) Maximizing the spread of influence through a social network. In: KDD'03 137–146
7. Kimura M, Saito K (2006) Tractable models for information diffusion in social networks. In: PKDD'06 259–271
8. Kimura M, Saito K, Nakano R (2007) Extracting influential nodes for information diffusion on a social network. In: AAAI'07 1371–1376
9. Kimura M, Saito K, Motoda H (2008) Minimizing the spread of contamination by blocking links in a network. In: AAAI'08 1175–1180
10. Leskovec J, Adamic L, Huberman B A (2006) The dynamics of viral marketing. In: EC'06 228–237
11. Newman M E J, Forrest S, Balthrop J (2002) Email networks and the spread of computer viruses. Phys Rev E 66:035101
12. Ng A Y, Zheng A X, Jordan M I (2001) Link analysis, eigenvectors and stability. In: IJCAI'01 903–901
13. Richardson M, Domingos P (2002) Mining knowledge-sharing sites for viral marketing. In: KDD'02 61–70
14. Saito K, Nakano R, Kimura M (2008) Prediction of information diffusion probabilities for independent cascade model. In: KES'08 67–75
15. Wasserman S, Faust K (1994) Social network analysis. Cambridge University Press, Cambridge, UK

# Meta-modeling the Cultural Behavior Using Timed Influence Nets

Faisal Mansoor[1], Abbas K. Zaidi[2], Lee Wagenhals[3] and Alexander H. Levis[4]

[1] fmansoor@gmail.com, George Mason University, Fairfax, VA
[2] szaidi2@gmu.edu, George Mason University, Fairfax, VA
[3] lwagenha@gmu.edu, George Mason University, Fairfax, VA
[4] alevis@gmu.edu, George Mason University, Fairfax, VA

**Abstract** A process that can be used to assist analysts in developing domain specific Timed Influence Nets (TIN) is presented. The process can be used to represent knowledge about a situation that includes descriptions of cultural behaviors and actions that may influence such behaviors. One of the main challenges in using TINs has been the difficulty in formulating them. Many Subject Matter Experts have difficulty in expressing their knowledge in the TIN representation. The ontology based meta modeling approach described in this paper provides potential assistance to these modelers so that they can quickly create new models for new situations and thus can spend more time doing analysis. The paper describes the theoretic concepts used and a process that leads to an automated TIN generation. A simple example is provided to illustrate the technique.

## 1 Introduction

The easy access to domain-specific information, often in some structured representation, and cost-effective availability of high computational power have provided analysts with unprecedented modeling and analysis capabilities in almost all areas of application, ranging from financial markets to regional and global politics. The analysis and decision problems often require modeling of subjective, informal, and uncertain concepts in a domain in order for an analyst to capture the required behavior of the domain. Several modeling and analysis formalisms now exist that try to address this need. The modeling of an uncertain domain using Bayesian Networks (BNs) is one of the most used of all such formalisms. The BN approach requires a subject matter expert (SME) to model the parameters of the domain – random variables – as nodes in a network. The arcs (or directed edges) in the network represent the direct dependency relationships between the random variables. The nodes in a BN and their interdependencies may represent the inter effects between political, military, economic, social, infrastructure, and information (PMESII) factors present in an area of interest. The strengths of these dependencies are captured as conditional probabilities associated with the connected nodes in a BN. A translation of the PMESII factors and the inter effects into the BN's analytical representation is a complex and time consuming task for the following two reasons: (a) experts knowledgeable in PMESII aspects of a domain often are not familiar with (or trained in) the analytical representation of BNs; (b) a complete specification of a BN requires an exponentially large number of parameter values (i.e., conditional probabilities) before it can be used for analysis and decision making problems.

Influence Networks [1] are a variant of BNs that address the latter problem with the BNs. They provide an intuitive and approximate language to elicit the large number of BN parameters from a very small set of inputs. Influence Nets are especially useful for modeling situations in which it is difficult to fully specify all parameter values required for a BN and/or where their estimates are subjective e.g., when modeling potential human reactions and beliefs. Wagenhals et al. [2] have added a special set of temporal constructs to the basic formalism of Influence Nets. The Influence Nets with these additional temporal constructs are called Timed Influence Nets (TINs). TINs have been experimentally used in the area of Effects Based Operations (EBOs) for evaluating alternative courses of actions and their effectiveness to mission objectives in a variety of domains, e.g., war games [3] , and coalition peace operations. A number of analysis tools have been developed over the years for TIN models to help an analyst in solving problems of interest [4-6].

The lack of familiarity with these analytical representations (i.e., BNs and/or TINs) prevents most domain experts and analysts from developing such models on their own and using them for the analysis tasks assigned. The tools implementing [7,8] some of these formalisms require prior knowledge and experience in modeling and, therefore, do not provide any assistance to such users. There is, however, a growing community of analysts who makes use of these analytical and quantitative formalisms resulting in a small, but expanding, repository of models addressing different PMESII aspects of a domain. There is, therefore, a need not only to facilitate the model building task, but also to utilize the existing models for building quick, larger and better domain models even without the need of domain experts.

This paper introduces Template Timed Influence Nets that will allow a domain expert (SME) to model an entire class of TINs addressing a problem using a compact representation. Template TINs model the problem at a generic level using abstract entities characterizing the problem domain. This model can then be instantiated for a particular situation by substituting abstract entities with concrete instances characterizing the situation.

A Template TIN also provides an effective solution for TIN reuse; it simplifies the TIN construction process by providing SME information about what to look for while developing a TIN. However, exploring available knowledge bases for information required for instantiating a Template TIN is also a complex and time-consuming task. Previously, the lack of machine understandable knowledge bases mandated manual data exploration, but with increasing popularity and use of structured knowledge representation and reasoning tools, it is now possible to automate the data exploration and Template TIN instantiation process. In this work, we are using ontology as the knowledge representation and reuse formalism. The choice of ontology is not arbitrary. Several proprietary and open-domain ontologies such as CYC [9], ThoughTreasure [10] and WordNet [11] have become available. Swoogle [12], an ontology search engine, has indexed more than 10,000 ontologies. Furthermore, several ontology construction and reasoning tools like Protégé [13], Swoop [14], Pellet [15], and ontology language standards like OWL [16] and RDF [17] make ontology an ideal candidate for knowledge representation and reuse.

By using Template TIN models along with ontologies, we can automate TIN construction, enabling a SME to develop comprehensive TINs along with saving time and effort. However, in order to instantiate a Template TIN from information available in an ontology, we will need some kind of mapping scheme that provides a definition of abstract concepts present in Template TIN in terms of abstract concepts and relations available in an ontology. In this paper, we describe how an ontology query language can be used to provide such a mapping.

The rest of the paper is organized as follows. In section II, we present the architecture of the developed Ontology Based TIN construction approach. Section III presents how this approach can aid a SME in the development of cultural behavioral analysis models. Section IV concludes the paper with a discussion on future research directions.

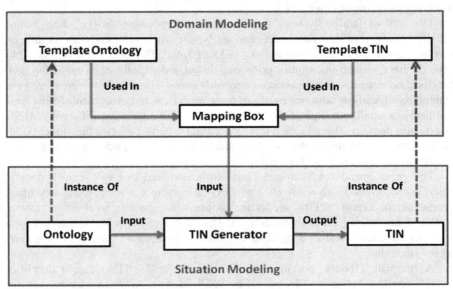

**Figure 1**: Architecture

# 2 Architecture

Figure 1 shows the architecture of the presented Ontology Based TIN Construction approach.

Table 1 describes the individual components and their role in the TIN construction of the architecture shown in Fig. 1.

**Table 1**. Components of Ontology Based TIN construction architecture.

| Component | Description |
|---|---|
| Ontology | Ontology is an explicit conceptualization of some domain of discourse. We can define ontology as a knowledge base composed of Terminology Box (TBox) and Assertion Box (ABox), where:<br><br>1. TBox is a finite set of concepts and a finite set of relations between the concepts.<br>2. ABox is a finite set of instances, relations between instances and relations between instances and concepts in TBox. |
| Template Ontology | A Template Ontology is essentially an ontology stripped of its ABox. It only contains TBox i.e. abstract concepts and relations between them. |
| Template TIN | Template TIN is an abstract TIN. Unlike a TIN, which only models a situation, a Template TIN can model a class of situations using a compact representation. |
| Mapping Box | Mapping Box defines Influences present in TIN in terms of concepts and relation available in Template Ontology. Specifically, Mapping Box is a set of mappings where each mapping is defined as a pair consisting of a Template Influence and an ontology query, the query establishes the link between Template Influence and ontology. |
| TIN Generator | Given an ontology describing a particular situation, TIN Generator uses the abstract definitions available in Mapping Box to produce a TIN specialized for the situation described by the input ontology. |

As shown in Fig. 1, TIN construction is a two-phase process consisting of a Domain-Modeling phase and a Situation-Modeling phase. In the Domain-Modeling phase, ontology and TIN templates are used to develop a generalized mapping that can be applied to any ontology compatible with the Template Ontology. Domain-Modeling is a process done only once. When a Mapping Box is created, instantiating a

TIN from a given instance ontology describing a particular situation becomes a completely automated process.

The generation process is presented in Fig. 2 as the function **GenerateInfluenceNet**. **This function** takes as input a Mapping Box **MBox** along with an ontology **O** and returns a TIN corresponding to **O**. **GenerateInfluenceNet** make use of the **Generate-Influence** procedure that, given a template influence and record, substitutes all abstract concepts in *template_influences* with appropriate instances from the *record*.

---

Procedure **GenerateInfluenceNet(MBox, O)**

    Let *tin* be an empty TIN.

    For each mapping pair (*template_influence, query*) in **MBox**

        Execute *query* over **O** to get a record set *R*.

        For every *record* in *R*

            *influence* = **GenerateInfluence**(*template_influence, record*).

            Augment *tin* with *influence*.

    Return *tin*.

---

**Figure 2**: TIN generation process

The described TIN construction process has been implemented as part of the Pythia [18] suite of applications. The implemented software package takes as input (a) an ontology expressed in OWL [16], (b) mapping rules expressed in SPARQL [19], and (c) Template TIN developed using Pythia for instantiating TINs. Pellet [15] is used as the ontology reasoning and query engine.

# 3 Application

To illustrate the concepts described in this paper, we have taken a TIN that was developed by subject matter experts in 2005 for Course of Action analysis concerning a region in Iraq [20]. It took several weeks of effort to create and verify the model. Part of the TIN included aspects of conflict resolution between different cultural groups. We have developed a generalization of this conflict resolution aspect of the TIN using the domain modeling approach described in Section 2 and attempted to use that approach to create a candidate TIN for a different domain problem.

The first step was to use the knowledge developed from the Iraq based TIN to create a Template TIN (Table 2a). The nodes in this template represent abstract concepts derived by replacing instances from the Iraq based TIN. From the concepts defined in the template TIN, plus additional understanding of the general nature of conflict resolution, a Template Ontology (Table 2a) was created, then a Mapping Box

(Table 2b) was developed to link the two template models. Finally, the Template Ontology was populated with information about a different scenario involving Darfur that is experiencing a conflict similar in nature to that in Iraq with multiple ethnic groups (i.e, Arab_Abbala and Darforians) vying for influence, an external peace broker (i.e., African Union), etc. The Darfur instance ontology was provided to the software application that used it and the Mapping Box to generate a TIN specialized for the Darfur scenario. The construction of a new instance TIN required that the variables in the Template TIN be replaced by the values available in the instance ontology with the help of mapping rules. For example, the variable [?cgA] in the proposition "[?cgA] cooperate in setting up government" gets 'Arab_Abbala' as its value from the Darfur ontology used for the illustration. This generated TIN could be used by a SME or analyst for course of action analysis as well as a host of other analyses provided by the TIN suite of tools [4-6]. The SME/analyst does not have to start each problem from scratch but can use the provided TIN as a starting point. The TIN can be used to examine alternatives for brokering meetings and agreements between specific cultural groups as well as the effects of various information dissemination strategies. The TIN can provide temporal course of action analysis indicating how long it will take for events to cause desired or undesired effects. Table 2(a) and Table 2(b) show the inputs provided to the software and the generated TIN for the Darfur example.

**Table 2(a).** Ontology Based TIN Construction Application.

| Component | Description |
|---|---|
| Template Ontology | 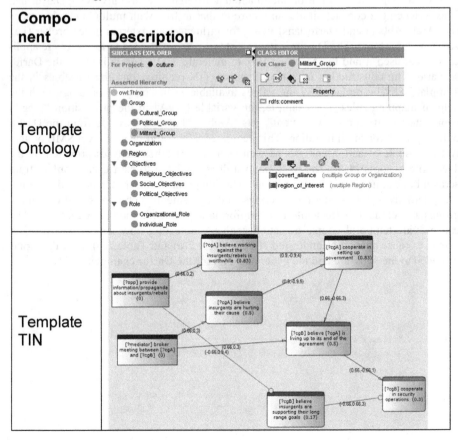 |
| Template TIN | |

**Table 2 (b).** Ontology Based TIN Construction Application.

| Component | Description |
|---|---|
| Mapping Box | Following is an example of a mapping rule, expressed in SPARQL, that provides a definition of the two types of cultural groups who are in conflict. The rule states that cgA is a group who is in majority while cgB is a group who is not in majority.<br><br>SELECT  ?cgA ?cgB<br>WHERE<br>{ ?cgA  rdf:type  this:Cultural_Group .<br> ?cgA  this:in_majority true.<br> ?cgB  rdf:type  this:Cultural_Group .<br> ?cgB this:in_majority false. } |
| Derived TIN | |

## 4 Conclusion

We have provided a brief description of an ontology-based approach to capture generalized knowledge about specific domain situations and use that knowledge to help analysts develop TINs quickly. The approach has been incorporated in a software package (Pythia) that can be used by SMEs as well as knowledge engineers. We believe the approach can lead to faster and better analysis of new complex situations that involve cultural factors that affect behavior of different groups including the analysis of alternative courses of action.

**Acknowledgement:** This work was supported by the US Air Force Office of Scientific Research under contract no. FA9550-05-1-0388

# References

1. Wagenhals, L. W. (2000) Course of Action Development and Evaluation Using Discrete Event System Models of Influence Nets. PhD Dissertation, George Mason University, Fairfax, VA.
2. Wagenhals, L. W. & Levis, A. H. (2000) Course of Action Development and Evaluation. In: Proceedings of the 2000 Command and Control Research and Technology Symposium.
3. Wagenhals, L. W. & Levis, A. H. (2002) Modeling Support of Effects-Based Operations in War Games. In: Proc. 2002 Command and Control Research and Technology Symposium, Monterey, CA.
4. Haider, S. & Zaidi, A. K. (2004) Transforming Timed Influence Nets into Time Sliced Bayesian Networks. In: Proceedings of Command and Control Research and Technology Symposium.
5. Haider, S. & Levis, A. H. (2005) Dynamic Influence Nets: An Extension of Timed Influence Nets for Modeling Dynamic Uncertain Situations. In: Proc. 10th Inter-national Command and Control Research and Technology Symposium, Washington DC.
6. Haider, S., Zaidi, A. K. & Levis, A. H. (2004) A Heuristic Approach for Best Set of Actions Determination in Influence Nets. In: Proc. IEEE International Conference on Information Reuse and Integration, Las Vegas.
7. Hudson, L. D., Ware, B. S., Mahoney, S. M. & Laskey, K. B. (2001) An Application of Bayesian Networks to Anti-Terrorism Risk Management for Military Planners. Department of Systems Engineering and Operations Research, George Mason University.
8. SIAM: Influence Net modeler (SIAC), http://www.inet.saic.com/inet-public/siam.htm.
9. Douglas, B. L. (1995). CYC: A Large-Scale Investment in Knowledge Infrastructure. Commun. ACM 38: 33-38.
10. Mueller, E. T. (1997). Natural Language Processing with ThoughtTreasure: Signiform.
11. George, A. M. (1995). WordNet: A Lexical Database for English. Commun. ACM 38: 39-41.
12. Ding, L., Finin, T., Joshi, A., Pan, R., Cost, R. S., Peng, Y., Reddivari, P., Doshi, V. C. & Sachs, J. (2004) Swoogle: A  Search and Metadata Engine for the Semantic Web. In: Proceedings of the Thirteenth ACM Conference on Information and Knowledge Management. ACM Press.
13. Grosso, W. E., Eriksson, H., Fergerson, R. W., Gennari, J. H., Tu, S. W. & Musen, M. A. (1999) Knowledge Modeling at the Millennium (the design and evolution of protege-2000). In: Proceedings of the Twelfth Workshop on Knowledge Acquisition, Modeling and Management, pp. 16-21.
14. Kalyanpur, A., Parsia, B., Sirin, E., Grau, B. C. & Hendler, J. (2006). Swoop: A Web Ontology Editing Browser. Web Semantics: Science, Services and Agents on the World Wide Web 4: 144-153.
15. Evren, S., Bijan, P., Bernardo Cuenca, G., Aditya, K. & Yarden, K. (2007). Pellet: A practical OWL-DL reasoner. Web Semantics: Science, Services and Agents on the World Wide Web 5: 51-53.
16. Mcguinness, D. L. & van Harmelen, F. (2004) OWL Web Ontology Language Overview. World Wide Web Consortium.
17. Beckett, D. (2004) RDF/XML Syntax Specification (Revised).
18. Pythia: Timed Influence Net Modeler, http://sysarch.gmu.edu/main/software/. SAL-GMU.
19. (2008) SPARQL Query Language for RDF.
20. Wagenhals, L. W. & Levis, A. H. (2007) Course of Action Analysis in a Cultural Landscape Using Influence Nets. Computational Intelligence in Security and Defense Applications, 2007. CISDA 2007. IEEE Symposium on pp. 116-123)

# A Validation Process for Predicting Stratagemical Behavior Patterns of Powerful Leaders in Conflict

Colleen L. Phillips[†], Stacey K. Sokoloff[†], John R. Crosscope[†],and Norman D. Geddes[†]

[†] (cphillips, ssokoloff, jcrosscope, ngeddes)@asinc.com, Applied Systems Intelligence, Inc., Alpharetta, GA

**Abstract.** As a subset of small group population beliefs, the emergent behaviors of powerful leaders in conflict with one another were modeled. Inputs to the model came from various news feeds that when ingested into the model, built evidence for the beliefs. The construction and execution of the leader belief model, a new predictive model validation process, and the implications are addressed.

## 1. Introduction

Persistent conflict is defined as a condition of protracted confrontation among state, non-state and individual actors that use a combination of harmful psychological and technological means applied to a society's local population to advance their strategic missions to achieve their political and ideological ends [1]. According to The official 2008 Army Posture Statement, America should expect to remain fully engaged throughout the world for the next several decades in persistent conflict [2]. Rogue nations have been seen to be dedicated to defeat their neighbors as nations and to eradicate them as a society through military and political powers. Trends such as access to WMD, emerging threats, and global connectivity do not exist in isolation [3]. They each interact with and complicate each other and the operating environment.

At West Point this year, the senior conference theme was `The Professional Military Ethic in an Era of Persistent conflict', which brought notice of the following questions regarding the implications of predictive modeling [4]:

- How will international, domestic, and technological trends shape the Army's future environment, and how will this environment affect the Army's roles and missions as well as the implications for knowledge, competencies, and values needed?
- What will the international, domestic, and technological strategic environments look like in the next 20 years, and can we begin to predict the rise of emerging powers given the advances in technology?

This paper presents a process for predicting plausible strategies that might be employed by powerful leaders when they use their political and military strengths to posture during a conflict between their countries. The 5 steps are outlined below:

H. Liu et al. (eds.), *Social Computing and Behavioral Modeling*,
DOI: 10.1007/978-1-4419-0056-2_20, © Springer Science + Business Media, LLC 2009

1) **Verifying and validating**, in a continuous manner, the predictive modeling process using a time-split historical perspective is discussed in section 2.

2) **Extracting grounded concepts of true beliefs** from leader's remarks obtained from multiple news feeds will be discussed in section 3. This process step assumes a tight coupling between reported remarks and actual beliefs and that the leader's and population's beliefs reflect the current state of the environment.

3) **Developing the belief inference models** of the leaders in conflict is discussed in section 4. The inputs, knowledge modules, and software suite work in unison to infer the believed political and military strength statistics. The belief statistics are predicted using the PreAct® Software Suite which employs a non-monotonic, abductive reasoner to make its inferences and has embedded Bayesian network computations to calculate the political and military strength statistics [5][6][7].

4) **Inferring the stratagemical behavior patterns.** The belief statistics generated over time are first categorized into belief states. Then these states are traced over time to form stratagemical behavior patterns and discussed in section 5.

5) **Refining the Model.** Continuous model refining, along with continuous system inputs will make the model richer, and slight changes in the model can be made to enable the importing of different state, non-state, and individual actor's beliefs.

## 2. The Belief Model Validation Process

Validation of the model's predictions was accomplished by considering a time-split historical case study. To test our model we took an evolving storyline from recent news. In August 2008, tensions between Georgia and Russia escalated into a full-blown military conflict after Georgian troops mounted an attack on separatist forces in South Ossetia. Using the Russia-Georgia conflict, tens of evidenced actions taken and stated by the countries' leaders and populations preceding a defined conflict event were used to populate the model and historical data to initialize. Then, the model was run on the initialized data and asked to predict the potential consequences (in terms of military and political strengths) of the actions leading up to six known, pre-defined conflict events. The modeling was done for the first three events, and both inputs and outcomes were incorporated. Next, we used a subject matter expert (SME) to predict what the perceived military and political strengths of both countries would be given tens of evidenced actions prior to a known, specific conflict event. Predictions were generated for the remaining three pre-defined (but un-modeled) events. Lastly, the rest of the inputs were modeled to obtain the calculated values of the perceived military and political strengths (Figures 1a and 1b).

| Modeled (MS) (PS) | | | Predicted by SME / Modeled (MS)(PS) | | |
|---|---|---|---|---|---|
| R-R: (A)(A) | (A)(A) | (A)(A) | (A/A)(A/A) | (A/A)(B/C) | (A/A)(A/C) |
| R-G: (B)(D) | (B)(D) | (A)(D) | (B/A)(D/D) | (B/A)(C/D) | (D/B)(D/D) |
| G-G: (A)(A) | (C)(B) | (A)(B) | (B/A)(A/C) | (C/A)(A/D) | (C/C)(B/D) |
| G-R: (A)(B) | (A)(B) | (A)(D) | (A/A)(D/C) | (A/A)(D/C) | (A/A)(C/D) |
| → E1 | → E2 | → E3 | → E4 | → E5 | → E6 → |
| 4/1/08 | 5/1/08 | 6/1/08 | 7/1/08 | 8/15/08 | 9/24/08 |

*Figure 1a: Timeline of events with Military and Political Strength (MS), (PS) values: (A (strong)=1-0.76; B=0.75-0.51; C=0.5-0.26; D(weak)=≤0.25). Key: Russia perceives Russia (R-R); Russia perceives Georgia (R-G); Georgia perceives Georgia (G-G); and Georgia perceives Russia (G-R).*

| 4/1/08 | E1: NATO refuses Georgia admission | 7/1/08 | E4: Russia accused of invading airspace |
|---|---|---|---|
| 5/1/08 | E2: Russia sends Peacekeepers | 8/15/08 | E5: Russia promises to pull-out troops |
| 6/1/08 | E3: Russia sends Railroad troops | 9/24/08 | E6: Russia sends troops to Ossetia |

*Figure 1b: The six significant events used for the modeling of Leader's Beliefs*

Public data sources and input from subject matter experts were employed to provide the storyline of six documented, significant, conflict events. The input was based on news accounts about the Russia-Georgia conflict with well known and documented outcomes, identified and encoded by the V&V team. The predictions were then compared to the actual values given by the model from actual evidenced events which followed from the previous action. The SME was correct 9/24 predictions, but not much reliability is given to this statistic given this was a subjective demonstration. Then the validation process begins again as the knowledge maps are revised based on the available, relevant data. The computational model is run again with on-going evidenced actions from where the model left off, repeating infinitely making it a truly a system that is both aware of its environment and predicatively sophisticated. It would also be interesting to predict events as well as strength statistics.

A metric could be derived to determine the similarity between predicted and observed events. Because the ultimate relevance of the predictions is their influence on tactical decisions, an effective metric should define similarity of predictions based on their relationships to the specific decisions which would result from each. A further challenge for the metric is that the output of the model's predictions will likely take the form of a set of outcomes with associated probabilities. Assuming that an observed event matches at least one of the possible predicted events, there is no way to determine whether the probability associated with the event was accurate. The only way of testing the probability distributions is to repeat the same test a number of times and count the number of times each outcome occurred. It will be important to modify the scenario to exhaust a significant number of variations, to confirm that the outcomes are sufficiently affected by the input.

## 3. The PMESII-PT Elements and the Operational Environment

The main technosocial resources that can be combined to affect campaign decision-making are seen in Figure 2a [8]. They are known as the PMESII-PT Elements – Political, Military, Economic, Social, Information, and Infrastructure with Physical and Time dimensions [9].   In order to understand our process, model, and its ultimate use, a description of all of the PMESII-PT resources or elements involved in the operational environment is needed. All of these elements can be manipulated by any military organization in order to reach their campaign objectives without relying strictly on military resources (the M in PMESII-PT). The first two of the seven factors were

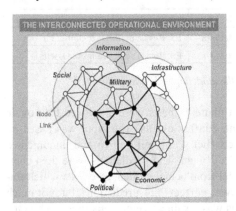 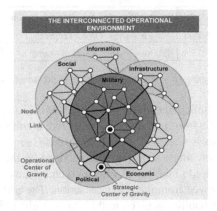

*Figure 2a (left): The PMESII-PT Elements of the Operational Environment. Source: Kem, 2007. Figure 2b (right): The PMESII-PT Elements showing how political and military actions can be the central theme of a stratagem. Source: Kem, 2007.*

chosen for this paper's analysis. Statistics for Political/Diplomatic and Military Strengths were computed (**PM**ESII-PT) using the leader belief model developed.

News feeds from 10 different countries were surveyed to collect inputs for our model in terms of triangulated, observable events covering the conflict between Russia and Georgia (Civil Georgia - http://www.civil.ge/eng/; Russian News and Information Agency - http://en.rian.ru/russia/20080313/; Russian Public Opinion Research Center - http://wciom.com/; Georgian Times - http://www.geotimes.ge; The St. Petersburg Times - http://www.sptimes.ru/index.php?i_number=1422; Georgia Today - http://www.georgiatoday.ge/articles.php?cat=Politics&version=392). These observed events were taken as grounded concepts, and knowledge maps were used to trace the effects of the events to the top level statistics. These statistics can assist the strategic planner in selecting the military operational strategy that manipulates beliefs in the desired manner. By enabling the military to effectively manipulate the PMESII-PT environment, through a collection of tactical strategies and plans, the integrity of the overall military campaign can be maintained [10]. The authors refer to this collection over time as a stratagemical behavior pattern [11]. An example is given in Figure 2b of a stratagem that calls for a show of military force (black node within the Military element) while performing a simultaneous, political action (additional black node within the Political element).

These patterns are an interconnected graph of nodes (operational, strategic actions to be taken) and links between the actions (plans and goals on how to manipulate their resources). How the graph changes over time and space gives an indication of the stratagemical behavior that emerges based on the chosen actions taken. The interaction between the nodes helps to identify key nodes, which are those nodes that are critical to the functioning of the campaign plan. This use of various PMESII-PT elements to control the perceived view of the operational environment is called stratagemical behavior – the use of stratagems. By manipulating the PMESII-PT elements, there is a possibility of adapting another group's view of the state of the PMESII-PT. A simultaneous set of actions could involve a political action (P) such as involving the leaders of "friendly" countries to sit down and talk with the Red Crescent about relief items needed in the devastated area, while the military action (M) could be to have the local governance stand watch in the area for snipers and bombers during the talks. The effects of using the PMESII-PT resources can be categorized into the strategic actions used.

## 4. The Powerful Leader Belief Model

The leader belief model is adapted from one developed for DARPA's Integrated Battle Command program [12]. This model was designed as a decision aid to assist military commanders in finding crisis resolution strategies when involved in future conflicts. Our specific model was developed to determine the influencing effect of each PMESII factor. The leader belief model only considered the P and M factors. The knowledge maps for determining political (P) and military (M) strengths are seen in Figures 3 and 4. PreAct® Software Suite is an intelligent agent framework that raises development to information and activity modeling, planning, and interpreting [5], [6], and [7]. Our current sociocultural computational modeling approach creates a dynamic model of each leader's beliefs and intentions and interprets them within a stratagem associate. These models are used with designs for incorporating the available PMESII-PT element inputs (in our case the P and M events via news sources) to plan actions that cause strategic effects that are desired or can be mitigated.

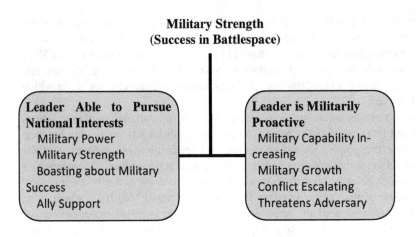

*Figure 3: The knowledge map for predicting Military Strength.*

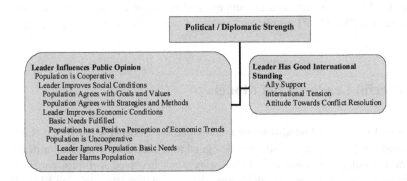

*Figure 4: The knowledge map for predicting Political/Diplomatic Strength.*

The outputs of the computational model were calculated Bayesian probabilities of a leader's perceived political or military strengths (MS, PS) with a value from 0-1 where 0 is weak and 1 is strong. These models have been made a part of the Stratagem Associate which the PreAct® cognitive engine interacts with. PreAct®, which is a mature and flexible framework for creating cognitive models, can interpret leader actions that change the underlying environment or that change relationships that can affect the beliefs and intentions of others, and hence, their underlying belief networks.

## 5. Technosocial Factors Affecting Powerful Leader Stratagemical Behavior Patterns

The knowledge maps containing relationships about the technosocial factors affecting the leader's belief of political and military strengths were formed into Bayesian belief networks and inputs from the news feeds were then used to populate the leader belief model. The computational model was run from all four perspectives of the leaders (Russia about Russia, Russia about Georgia, Georgia about Georgia, and Georgia about Russia). The results are placed in a state-space graph and presented in Figure 5.

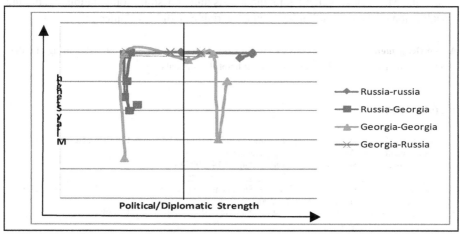

Figure 5:    Graph of the four perspectives of Military and Political Strengths.

If you divide the graph into 4 quadrants, they represent the four outcomes if you follow certain strategies in order to try and maintain yours and other's perceived strengths in a state 1-4 where state 1 is superior power. In terms of M and P: 1) upper right: dominant military and political power, 2) lower right: dominant political, but not military, 3) upper left: dominant military, but not political power, and in the 4) lower left: inferior power both in military and political strengths.   This representation can be traced over time and between significant events to form the stratagemical behavior pattern of a leader during conflict resolution. Eventually, after enough patterns are generated and analyzed statistically, predictions about leader's actions could be made on the basis of evidence gathered in the form of necessary input. Then, courses of actions could actually be planned, mitigated, predicted and strategies for belief persuasion and maintenance developed.

# 6. Implications and Conclusions

The continuous validation process developed was tested for a conflict scenario and proved to be useful in the absence of experimental or statistical analyses. The predictions are valid in the week's to month's timeframe, but decay for longer periods. Even chaotic weather patterns can be predicted for the next few days, but are not very reliable past that. The stratagemical behavior patterns emerged following events that affect the perceived military and political strengths of two leaders. Following these patterns over time will enhance the predictability of the model by showing the emerging trends of perceived beliefs. This was accomplished for only two of the PMESII factors, P and M. Looking at all combinations is quite possible, computationally, but the visualization of the interaction between all variables will be a challenge.

**Acknowledgements** The authors wish to thank Corey Canis for his hours of web surfing and data input generation and the SBP reviewers for their helpful comments.

# References

1. Lennox, MG R. (2007). Modeling and Simulation in an Era of Persistent Conflict. , Headquarters, Department of the Army G-3/5 Strategic Plans, Concepts, & Doctrine Division.
2. US Army Posture Statement. Retrieved on August 14, 2008 from http://www.army.mil/aps/08/information_papers/prepare/Persistent_Conflict.html.
3. Geren, P., Gen. Casey, G.W., Jr. (2008). A Statement on the Posture of the united States Army 2008). 2nd Session, 110th congress, Feb. 26, 2008.
4. West Point Senior Conference XLV. West Point, NY. Retrieved on July 4, 2008 from (http://www.dean.usma.edu/departments/sosh/seniorconf/.
5. Geddes, N.D. and Atkinson, M.L. (2008). An approach to modeling group behaviors and beliefs in conflict situations. In Social Computing, Behavioral Modeling, and Prediction. Eds. H. Liu, J.J. Salerno, M.J. Young; Springer, New York, pp. 46-56.
6. Geddes, N.D. (1997). Large scale models of cooperative and hostile intentions. Proceedings of IEEE Computer Society International Conference and Workshop (ECBS'97), Monterey, CA.
7. Geddes, N.D. (1994). A model for intent interpretation for multiple agents with conflicts. Proceedings of the IEEE International Conference on Systems, Man and Cybernetics, SMC-94. San Antonio, TX.
8. Kem, J.D. Colonel (2007). Understanding the Operation Environment: The Expansion of DIME. April-June. University of Military Intelligence.
9. Mattis, Gen. J.N. (2008). USJCOM Commander's Guidance for Effects-Based Operations. Memorandum for US Joint Forces Command. August 14, 2008.
10. Phillips, C.L., Geddes, N., and Kanareykin, S. (2008). A Balanced Approach for LLOs Using Group Dynamics for COIN Efficacy. Proceedings of the 2nd International Applied Human Factors and Ergonomics Conference, July 14-17, Las Vegas NV.
11. Phillips, C.L., Geddes, N., and Crosscope, J. (2008). Bayesian Modeling using Belief Nets of Perceived Threat Levels Affected by Stratagemical Behavior Pattern. Proceedings of the 2nd International Conference on Cultural Computational Dynamics, Sept 15-16, Washington, D.C.
12. Dyer, D. E. (2004). The Reflex PMESII Model. Active Computing. Retrieved on August 14, 2008 from http://www.activecomputing.org/papers.html.

# Control of Opinions in an Ideologically Homogeneous Population

Bruce Rogers[1] and David Murillo[2]

[1]rogers@mathpost.asu.edu, Arizona State University, Tempe, AZ
[2]dlm35@mathpost.asu.edu, Arizona State University, Tempe, AZ

**Abstract** We describe a simple model for the process of changing opinions where clusters of different opinions form in a society. We are interested in what happens to these clusters after they are formed. Is the cluster robust to the introduction of people with differing opinions? We focus on the case where some people are influenced more easily than others and examine the impacts of network structure and different strategies aimed at changing the mean opinion of the population. We find that the more variation in a population's open-mindedness, the easier it is for the population to be influenced, and exploiting the structure of the network allows the society influenced more efficiently

## 1 Introduction

Members of a society often have a wide range of opinions on a wide range of subjects. While these opinions often change in time due various factors including changing information and changing preferences, they can also be directly influenced by the opinions of others. We offer a simple model of opinion change in a population where opinions are influenced soley due to the interactions of other people and their opinions. Suppose there are $N$ members of the population, agents, and the opinion of each agent $i$ at time $t$ is represented with a real number $x_t(i) \in \mathbb{R}$. One way to update the agents' opinion in time is for each individual to average its opinion with the opinions of a group of its neighbors. Then, under some weak conditions, the agents opinions converge asymptotically to a single consensus value [9]. Algorithms such as this for reaching consensus have recently received a lot of attention with applications in sensor networks, robotics, flocking and many others[3, 10, 6].

However, total consensus is rarely achieved in a society of more than just a few actors. By modifying the amount of information each agent will accept from the society, we can capture the fragmentation of opinions. Suppose each agent has an opinion $x_t(i)$ taking a value between 0 and 1. Instead of averaging its opinion with all its neighbors, each agent will consider only those opinions close to its own while ignoring all the others. More formally, suppose $\varepsilon > 0$, then agent $i$ will ignore the opinion of agent $j$ if their opinions are more than $\varepsilon$ apart. We will refer to $\varepsilon$ as the *bound of confidence* or the *confidence radius* [4, 5, 8]. The constant $\varepsilon$ can be viewed as a measure of an agent's open-mindedness.

H. Liu et al. (eds.), *Social Computing and Behavioral Modeling*,
DOI: 10.1007/978-1-4419-0056-2_21, © Springer Science + Business Media, LLC 2009

The set of confidants of agent $i$ at time $t$ is given by the set $N_t(i) = \{j \; i : |x_t(i) - x_t(j)|\} < \varepsilon$. Then we can define the change in opinions over time by

$$x_{t+1}(i) = \frac{1}{|N_t(i)|} \sum_{j \in N_t(i)} x_t(j) \qquad (1)$$

Suppose each agent is given an initial opinion $x_0(i)$ uniformly at random in the unit interval $[0,1]$. As the agents update their opinions Equation 1, the agents start to cluster around certain opinions. In a finite amount of time, the population will segregate itself into some number of opinion clusters [7]. The number of clusters that form depends on the radius of confidence $\varepsilon$, the specific initial opinions and the social network.

The existing literature on this social learning model concentrates on the process of cluster formation from random initial opinions [2, 11, 8]. However, we wish to consider how to persuade opinion clusters once they have formed. Suppose the system has reached a steady state where the $N$ agents in a population have coalesced to some opinion clusters. How can we change the opinions of these actors? If we introduce a new group of agents with different opinions, how quickly will the original population change its (average) opinion?

For simplicity we concentrate on the scenario where there is only a single opinion cluster. Without loss of generality, suppose this opinion is at the origin: $x_t(i) = 0$, for all $i$. We wish to examine robustness of the mean opinion to change. Since the population is at equilibrium, we must either change the opinions of some of the agents or add new agents to the system.

We suppose there are some agents whose behavior we can control and call them *special agents*. We focus on the situation where there are only a small number special agents relative to the size of the population. The special agents have $\varepsilon = 0$ so that they will not change their opinion, and their initial opinions differ from 0, $x_s(0) \neq 0$. The opinions of the special agents will act as a driving force, pulling the opinion of the population toward the opinion of the special agents. We then consider a few basic questions: If we have only a few special agents at our disposal, how can we move the original population's mean opinion quickly? How does the structure of the social network and the distribution of $\varepsilon$ affect the change in the mean opinion of the population?

In Section 2, a complete social network is considered, and in Section 3 the social network is modeled as a scale-free network. In each section, we consider two different distributions of the open-mindedness of the population: first a uniform distribution and then a normal distribution. We see that the distribution of $\varepsilon$ on the population heavily influences the mean opinion of the population.

## 2 Complete Network

Suppose there is a population of $N = 1000$ actors all with initial opinion $x_0(i) = 0$. We add to the population $K = 10$ special agents with opinion $a = .9$. Again, the special agents are *not influenced* by the rest of the population; they have an $\varepsilon$ of 0. So, the total number of agents is 1010, and in this section we assume that every agent in the social network is connected to every other agent (thus we have a complete network).

The population updates its opinion according to Equation 1. If all the agents have $\varepsilon < 0.9 = a$, then clearly there is no change to their initial opinions. On the other hand, if all the agents have $\varepsilon > 0.9$, the population opinion converges geometrically to the special agents' opinion. In the next two subsections, we explore what effect different distributions of open-mindedness has on the opinion dynamics.

### 2.1 Uniformly Distributed $\varepsilon$

Suppose the values of $\varepsilon$ are spaced evenly such that they follow the uniform distribution. There are 1000 agents in the population, and we run simulations with agent $i$ given $\varepsilon_i = \frac{i}{1000}$. The simulation are run for 1000 time steps. The upper-left of Figure 1 plots the mean opinion of the population over time. Except for 2 agents with the smallest values of $\varepsilon$, the entire population drifts toward the opinion of the special agents. We would expect including more special agents would force the mean opinion to converge more rapidly and this is true, initially. If more than 500 special agents are added, the population fragments with a proportion heading toward the special agents and the remainder going to an equilibrium opinion bounded away from the special agents' opinion, a.

As $t \to \infty$ the agents' opinions stop changing, so the mean opinion goes to a constant. However, not every opinion converges to the mean. For 10 special agents, less than 1% of the population has opinion significantly lower than the special agent opinion. Adding more special agents can actually increase the stratification of societal opinions, but the amount is negligible until the special agents comprise about one-third of the total population. Since there are so many special agents, the opinions of those agents that are open-minded to the special agents converge too quickly and are not able to influence the less open-minded agents. This phenomenon of stratification is not observed when $\varepsilon$ in normally distributed, as in the next subsection.

### 2.2 Normally distributed $\varepsilon$

Here we suppose that the values of $\varepsilon$ are spaced evenly such that they follow a normal distribution with mean 0.5 and standard deviation 0.125. As before, we take a population with 1000 agents and 10 special agents and run the simulations until

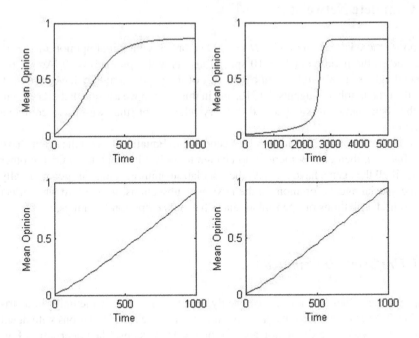

**Fig. 1** Mean opinion of 1000 agents with complete social network over time. Upper Left: Level of open-mindedness, $\varepsilon$, is uniformly distributed over population, and special agents have fixed opinion 0.9. Upper Right: $\varepsilon$ normally distributed with mean 0.5 and standard deviation 0.125; special agents have fixed opinion 0.85. Lower Left: Uniformly distributed $\varepsilon$; special agents start with opinion 0.1 and move toward 1. Lower Right: Normally distributed $\varepsilon$ and special agents same as lower left.

the mean opinion approaches equilibrium. If the special agents all are given opinion 0.9, there are

The upper-right of Figure 1 plots the mean opinion over time when the special agents all have opinion 0.85; notice that the time scale is different than in the case of uniformly distributed $\varepsilon$. When special agents comprise about 1% of the population, the agents with normally distributed $\varepsilon$ take much longer to converge to equilibrium than a population with uniformly distributed $\varepsilon$. Since the mean opinion changes more slowly, more of the population is captured and we avoid the stratification that occurs with uniformly distributed $\varepsilon$.

### 2.2.1 Changing Strategy

Previously, our strategy was to have the special agents' opinions fixed at the edge of the distribution of $\varepsilon$. In that case, a population with normally distributed $\varepsilon$ takes 3 times longer to converge to the special agent opinion than a population with uniformly distributed $\varepsilon$, see Figure 1. In order to improve the rate of convergence, we now place the special agents' opinion closer to that of the population's and have the special agents periodically increase their opinion.

Even though it is not modeled in the dynamics, an actual population may become less inclined to heed the opinion of others who continually change their opinion, so the special agents start with opinion 0.1 and keep that opinion for 100 time steps. Then they assume an opinion of 0.2 for 100 time steps and then change their opinion to 0.3. The special agents continue to increase their opinion by 0.1 every 100 time steps until they reach the maximum opinion of 1.

As before, we start with a population size of 1000 having initial opinion 0 and add 10 special agents who behave as described in the above paragraph. The simulations are run for 1000 time steps. The bottom of Figure 1 shows the population's mean opinion over time for this new strategy. On the left is a population with uniformly distributed $\varepsilon$, and the right has normally distributed $\varepsilon$. Under this new strategy, both populations enjoy an approximately constant rate of change. This is a marked improvement in performance over the previous case for a population with normally distributed $\varepsilon$. However, when $\varepsilon$ is uniformly distributed, the population opinion converges faster if the special agent opinions are fixed near the edge of the distribution.

# 3 Scale Free Network

Until now, we have supposed that every agent has the ability to interact with every other agent, and the flow of information is impeded only by the radius of an agent's confidence $\varepsilon_i$. However, large populations are often structured in a more restrictive communication network. In this section we consider a population with a scale free communication network constructed by preferential attachment [1]. The graph is random, but we use the *same network* for every run of the simulations.

So, we fix a random scale-free network $G$ with 600 nodes to represent the population. As before, each member of the population has initial opinion $x_0(i) = 0$ and some radius of confidence $\varepsilon_i$. Then we pick $K$ existing nodes at random to convert to special agents. If the entire population has the same level of open-mindedness, say $\varepsilon = 1$, the opinion dynamics behave similarly to a population with a complete social network. The entire population converges to the special agents opinion, albeit more slowly as it takes longer for information from the special agents to seep through the network. As the number of special agents increases, their opinion is able to spread throughout the network more quickly.

## 3.1 Uniformly Distributed ε

In this section we fix the number of special agents to $K = 10$. For a population of 600 embedded in the scale free network, we sample $\varepsilon_i$ uniformly from the unit interval. The special agents are picked at random from the population and given the fixed opinion a=0.9.

The upper-left of Figure 2 shows the population's mean opinion upon reaching equilibrium for 50 different runs of the simulation. In every run, part of the population moves toward the special agents' opinion. However, some part of the population is always left with opinion 0. We see that randomly picking special agents with a fixed opinion does have an affect on the population's mean opinion, but it will not necessarily converge to the special agents' opinion. In 12% of the cases, the population's mean opinion does not progress past 0.5.

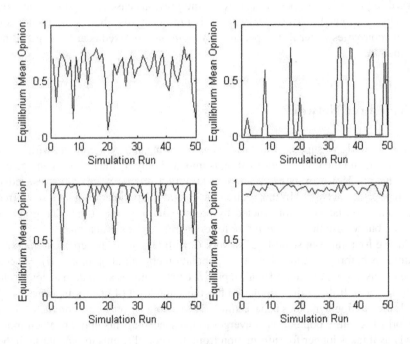

**Fig. 2** Equilibrium mean opinion of a population with scale-free communication network for 50 simulation runs. Left: Uniformly distributed ε. Right: Normally distributed ε. Upper: Special agents with fixed opinions. Lower: Special agents start with opinion .1 and move toward 1.

The lower-left of Figure 2 shows the results of 50 simulation runs when the special agents employ the strategy of Section 2.2.1. Now, the special agents are able to influence a much larger proportion of the population, and the equilibrium mean opinion increases significantly from the case with fixed opinion special agents.

However, we still see that if the special agents are poorly placed in the network, surrounded by very few neighbors with small $\varepsilon$, as much as 50% of the population will not feel the influence of the special agents.

## 3.2 Normally Distributed $\varepsilon$

We now consider the same situation with 10 special agents picked randomly from the nodes of the scale-free network, except the $\varepsilon_i$ are now normally distributed with mean 0.5 and variance 0.25. We truncate the distribution so $\varepsilon \geq 0.1$. The plots on the right of Figure 2 show the mean opinion for 50 simulation runs employing both special agent strategies. In the top-right, the 10 special agents have fixed opinion 0.85. Only 11 times out of 50 do any members of the population change their opinion. And of these 11 times, only nine times was the final mean opinion greater than 0.5 (whereas with uniformly distributed $\varepsilon$ the final mean opinion was only less than 0.5 6 times). A single agent has only probability 0.0026 of having $\varepsilon$ greater than 0.85, so we expect that only 1 or 2 agents in the population can be influenced by the special agents. If instead we fix the special agent opinions at 0.8, we expect about 5 members of the population to have $\varepsilon$ this large. Then the probability of the special agents influencing the population increases to about 40%.

The lower-right of Figure 2 shows the affect of the special agents employing the strategy of Section 2.2.1. In every run, the population's mean opinion increases to greater than 0.9. A very small number of agents are still left at opinion 0, but this strategy is quite successful even when the special agents are on low degree nodes.

## 3.3 Targeting High Degree Nodes

If we have no knowledge of the communication network, we must pick the special agents randomly from the population. In this section we assume that the network structure is known, and we target the nodes with the highest degree to become special agents. We use the same randomly generated scale-free network as in the previous sections. This network has 600 vertices with mean degree 3, and the two highest degrees are 66 and 54. Previously, we have chosen 10 special agents at random from the vertices. Now, we will choose only one special agent, but it will be either the vertex with highest degree or second highest degree.

Figure 3 shows the results for choosing a single high degree node (K=1) to be a special agent. In all the simulations, the population is given initial opinion 0, and the special agent has fixed opinion 0.85. The left two graphs have a population with uniformly distributed $\varepsilon_i$, and the graphs are the right are for normally distributed $\varepsilon_i$. For the upper graphs, the special agent is chosen to be the vertex with highest degree, and the special agent is the second highest degree vertex on the bottom.

At first glance, there does not seem to be an appreciable difference from this situation and the fixed agent strategy in the upper plots of Figure 2. However, please note that the plots in Figure 2 employed *ten times* as many special agents. That is to say, knowledge of the network structure and ability to exploit high degree nodes remarkably increases the special agents' influence on the population.

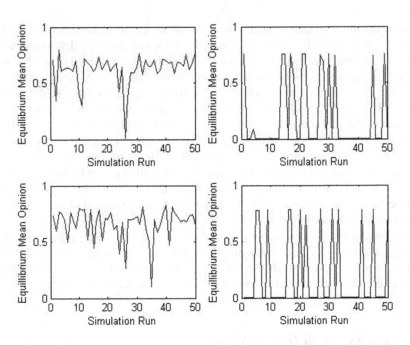

**Fig. 3** Equilibrium mean opinion of a population with scale-free communication network for 50 simulation runs. Left: Uniformly distributed $\varepsilon$. Right: Normally distributed $\varepsilon$. Upper: A single special agent with fixed opinion on the highest degree vertex. Lower: A single special agent on the 2nd highest degree vertex.

## 4 Conclusion

We have provided a simple model of opinion change in a population where opinions are taken to be real numbers and the change in opinions is given by Equation 1. Each agent $i$ in the population has a radius of confidence $\varepsilon_i$ which models its willingness to change its opinion. Then we studied the dynamics of the mean opinion under two different strategies, a static one and one when the special agent's opinion increases incrementally.

If the communication network between agents is sparse, it is possible that adding special agents to a population has only limited effect on the population's mean opinion. Essentially, if the special agents are not well connected, the population ignores them. We have determined two ways to increase the influence of the special agents. The first is to pretend they differ from the population opinion by only a small amount. Then they methodically increase their opinion. This strategy does not rely on knowledge of the communication network. If the communication network is known, controlling the high degree nodes also significantly increases the probability of influencing the population.

The simplicity of our model allows detailed analysis, and many extensions are possible. We assume that an agent weighs all the opinions of its confidants equally, and it would be straightforward to give some agents more influence than others. However, actual opinions and ideas are much too complex to be captured by an averaging process in a Euclidean space. To model real-world opinion formation, it will be necessary to include more accurate cognitive processes in the model.

# References

1. Barabasi, A. and Albert, R. 1999. "Emergence of Scaling in Random Networks," *Science*, vol. 286, pp 509-512.
2. Blondel, V., Hendrickx, J., and Tsitsiklis, J. "On the $2R$ Conjecture for Multi-agent Systems."
3. Boyd, Ghosh, Prabhakar, and Shah. "Gossip algorithms: design, analysis, and applications." In *Proceedings of IEEE Infocom*, Miami, March 2005.
4. Deffuant, G., Neau, D., Amblard, F. and Weisbuch, G. "Mixing Beliefs among Interacting Agents." Advances in Complex Systems, 3, pp 87-98, 2000.
5. Hegselmann, Rainer and Krause, Ulrich. "Opinion Dynamics and Bounded Confidence: Models, Analysis and Simulation," Journal of Artificial Societies and Social Simulation, vol. 5(2), 2002.
6. Jadbabaie, A., Lin, J., and Morse, A. 2003. "Coordination of Groups of Mobile Autonomous Agents Using Nearest Neighbor Rules," *IEEE Transactions on Automatic Control.* vol. 48, no. 3 pp 988-1001.
7. Lorenz, Jan. "A stabilization theorem for dynamics of continuous opinions." Physica A, vol. 355, pp 217-223, 2005.
8. Lorenz, Jan. "Continuous Opinion Dynamics under Bounded Confidence: A Survey." arXiv:0707.176v1, 12 July 2007.
9. Tsitsiklis and Bertsekas, *Parallel and Distributed Computataion*, Prentice Hall, 1989.
10. Vicsek, Czirok, Ben-Jacob, Cohen, and Shochet, "Novel Phase Transition in a System of Self-Driven Particles," Physical Review Letters, vol. 75, no. 6, 1995.
11. Weisbuch, G, Deffuant, G, Amblard F, and Nadal, JP. 2002. "Meet, Discuss, and Segregate!" *Complexity*, vol. 7, no. 2, pp 55-63.

# Estimating Cyclic and Geospatial Effects of Alcohol Usage in a Social Network Directed Graph Model

Yasmin H. Said[1] and Edward J. Wegman[2]

[1]yasid99@hotmail.com, Gorge Mason University, Fairfax, VA
[2]ewegman@gmail.com, George Mason University, Fairfax, VA

## 1 Introduction

Alcohol use and abuse contributes to both acute and chronic negative health out-comes and represents a major source of mortality and morbidity in the world as a whole (Ezzati et al., 2002) and in the developed world, such as the United States, where alcohol consumption is one of the primary risk factors for the burden of dis-ease. It ranks as a leading risk factor after tobacco for all disease and premature mortality in the United States (Rehm et al., 2003; Said and Wegman, 2007). In a companion piece to this paper, Wegman and Said (2009) outline a graph-theoretic agent based simulation tool that accommodates the temporal and geospatial dimen-sions of acute outcomes. In order to complement that modeling paper, this paper focuses on methods to exploit temporal and spatial data with the idea of calibrating the model to include these effects. The model developed in Wegman and Said (2009) incorporates a social network component. The overall goal of the model is not just to simulate known data, but to provide a policy tool that would allow decision mak-ers to examine the feasibility of alcohol-related interventions. Such interventions are designed to reduce one or more acute outcomes such as assault, murder, suicide, sexual assault, domestic violence, child abuse, and DWI-related injuries and deaths. By adjusting conditional probabilities, the effect of interventions can be explored without actually introducing major societal policy actions.

## 2 Estimating Temporal and Cyclic Effects

It is well known that there are substantial seasonal and temporal effects associated with alcohol use and its acute outcomes (Fitzgerald and Mulford 1984; Cho et al. 2001; and Carpenter 2003). For purposes of analysis of interventions, it is desirable to understand when and where interventions may be most effective. Alcohol use

H. Liu et al. (eds.), *Social Computing and Behavioral Modeling*,
DOI: 10.1007/978-1-4419-0056-2_22, © Springer Science + Business Media, LLC 2009

shows patterns on multiple scales including time-of-day, day-of-week, and week-of-year as well as month-of-year. The detailed construction of probabilities conditioned on job class, ethnicity, gender, socioeconomic status, residential location, age, and a host of other demographic factors results in many thousands of conditional probabilities to be estimated. Adding temporal effects specific to each of the many thousands of combinations of relevant variables and acute outcomes is unrealistic with existing data sets. In order to incorporate temporal effects in the model outlined in Wegman and Said (2009), we need to develop proxies for the temporal effects. The first part of this analysis examines data from the Virginia Department of Motor Vehicles alcohol-related fatal crashes for the period 2000-2005 (896,574 instances). After summarization, the data set has 2,192 instances (365×6). The date of the crash is used to extract the year, month, week, and day of the week. Analysis is done on the number of crashes on a particular day of the week. The second part of our analysis focuses specifically on Fairfax County.

The main data resource consists of state-wide data from Virginia Department of Motor Vehicles on all alcohol-related crashes from 2000 to 2005. These data were made available with time-of-day as well as date information. Based on the date information, we are able to infer day-of-week and month-of-year information. These data are aggregated for each year and plotted in Figure 1 by time-of-day. There is remarkable consistency in these plots suggesting that there is comparatively little yearly effect. As one might expect, drinking behaviors increase during late afternoon and early evening and this is reflected in the increases in alcohol-related fatal crashes. The very early morning events (1 to 3 a.m.) after closing hours of most alcohol outlets are associated with attempts to drive after an extended evening of drinking.

Figure 2 plots the crashes by day-of-week. Again, the data are aggregated across the years by day of the week. There is once again remarkable consistency over the annual period. As one might reasonably expect, drinking behaviors increase with the onset of the weekend. Typically, drinking behaviors begin on Friday, are most intense on Saturday, and begin to subside on Sunday.

Examining the month-of-year effect, we find that the lowest crash rates are in February, March, and April. The traditional holiday period, October, November, and December incur the greatest number of fatal crash events. Surprisingly, the peak of the summer driving/summer holiday season is comparatively low. Even more surprising is the relative peak in May. We conjecture that this is associated with the end of the semester for high schools and colleges with proms, graduations, and other social events for high schools and collegiate celebrations at the end of final exams and commencements.

Figures 1-3 suggest that there are very substantial time-of-day effects, substantial day-of-week effects, a somewhat less pronounced monthly effects, and virtually no variation over the years. In order to investigate this more analytically, we consider mixed effects linear models. The response variable, alcohol-related fatal crashes, is skewed to the left. This indicates non-normality. The square root transformation is used for this data set. We consider the mixed effects model: $y_{ijk} = \mu + \alpha_i + \beta_j + \gamma_k + \varepsilon_{ijk}$ where $y_{ijk}$ is the observation of the $i$th day of the $j$th week of the $k$th year. In this

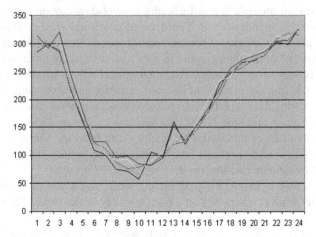

**Fig. 1** Alcohol-related crashes by time-of-day.

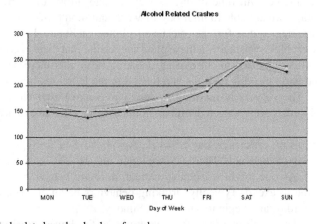

**Fig. 2** Alcohol-related crashes by day-of-week.

case, $i = 1, \ldots, 7$, $j = 1, \ldots, 52$, and $k = 1, \ldots, 6$ where $\alpha_i$ is the day-of-week effect, $\beta_j$ is the week-of-year effect, $\gamma_k$ is year, $\mu$ is the fixed intercept, and $\varepsilon_{ijk}$ is the noise. Day-of-week variations are highly significant, week-of-year variations marginally significant, and the year effect is not significant.

While the alcohol-related crash data are state-wide data, they do not capture temporal characteristics of DWI arrests and alcohol-related hospital admissions. In Figure 4, we have the boxplot of DWI arrests by day-of-week. This strongly resembles the day-of-week plot for alcohol-related fatal crashes state-wide. Of course, there are many more DWI arrests than there are alcohol-related fatal crashes. In these figures, day 1 is Monday. Both fatal crashes and DWI arrests are lowest on Tuesdays. Figure 5 is the boxplot of alcohol-related treatment admissions. This essentially is

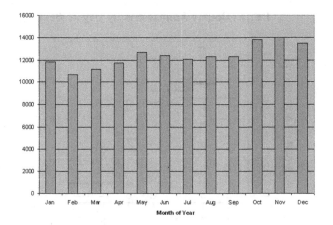

**Fig. 3** Alcohol-related crashes by month-of-year.

## Fairfax County DWI by Day of Week

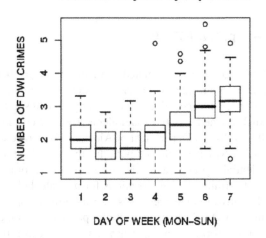

**DAY OF WEEK (MON–SUN)**

**Fig. 4** Fairfax County DWI arrests by day-of-week.

inversely correlated with the DWI and crash data. Here, Tuesday is the peak day for admissions to treatment.

In summary, the time-of-day and day-of-week alcohol-related crashes provide an excellent proxy for the critical temporal effects needed in our model and allow us to incorporate these temporal effects.

**Fairfax County Admissions By Day of Week**

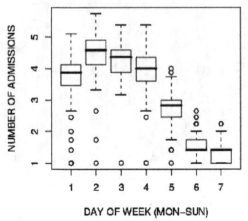

**Fig. 5** Fairfax County hospital admissions for alcohol treatment.

## 3 Estimating Geospatial Effects

Because a large percentage of alcohol-related incidents take place between alcohol outlets and the alcohol user's residence, either at the outlet itself where drinking behaviors take place, at the residence, which is the other principal site for alcohol use, or in between for such incidents as DWI and alcohol-related crashes, we take advantage of data collected by the Department of Alcoholic Beverage Control (ABC) in Virginia. The ABC survey is conducted periodically at ABC Stores. Customers are asked their residential zip code and this information is recorded for planning purposes in order to determine the best locations for new alcohol outlets. Presumably, best is judged relative to enhanced sales, not in terms of optimal societal or public health benefits. The ABC surveys essentially provide a two-mode social network and arranged in the form of an adjacency matrix, we can form associations between alcohol outlets and zip codes. A gray-scale rendition of the ABC by zip code adjacency matrix is given in Figure 6.

An interesting statistic associated with Figure 6 is that purchasers of alcohol in Fairfax County, Virginia have their residence as far away as 340 miles. There are a large number of purchasers of alcohol in Fairfax County that live approximately 100 miles away from Fairfax County. We conjecture that these are commuters whose local ABC stores would be closed by the time they arrive at their residences. Fairfax County is West of Washington, DC and would be on a commuter route for those living in the more Western parts of Virginia and West Virginia.

Still, if we focus on the 48 zip codes in Fairfax County and the 25 ABC outlets in Fairfax County, we can establish approximate commuter routes for purchasers of

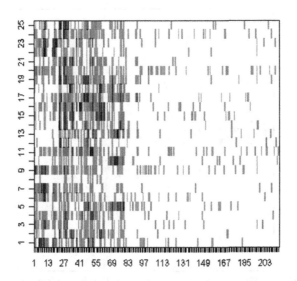

**Fig. 6** ABC outlets versus zip codes. There are 25 ABC outlets in Fairfax County, Virginia and more than 230 residential zip codes representing the homes of people who purchase distilled spirits in Fairfax County. Fairfax County contains 48 zip codes, so a considerable amount of alcohol is purchased in Fairfax County by non-county residents. In this Figure, the rows represent the alcohol outlets and the columns represent the residential zip codes.

alcohol and thus determine what may well be local hotspots for alcohol-related incidents, by virtue of the amount of alcohol purchased as well as the likely commuter routes. ABC stores are the only off-license establishments in Virginia that sell distilled spirits. Other off-license establishments as well as on-license establishments are not accounted for in this exercise. Still, the ABC stores are a reasonable proxy for geospatial effects. Figure 7 represents the map of Fairfax County with zip codes.

Based on the adjacency matrix of the two mode network weighted by the amount of alcohol purchased (in terms of dollar sales) and the centroids of the zip codes, we can estimate the distribution of distances traveled between outlet and residence and to at least a crude approximation, estimate the routes alcohol users take between outlet and residence. These are relatively crude proxies, but reasonable first approximations for purposes of calibrating our model.

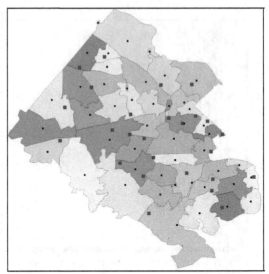

**Fig. 7** Fairfax County, Virginia with zip codes outlined. The shading is proportional to the actual number of acute events in each zip code. The larger (red) markers are the locations of ABC stores. The smaller (black) markers are the centroids of each zip code. From the two-mode adjacency matrix restricted to Fairfax County we can determine approximate distances and routes.

## 4 Conclusions

In this paper, we develop an estimation procedure for a comprehensive modeling framework based on agent-based simulation, social networks, and directed graphs that captures some of the complexity of the alcohol ecological system. We provide a framework for temporal and geospatial effects and discuss proxies for approximating these effects. To completely specify the framework outlined in Wegman and Said (2009), considerable additional data collection efforts must be undertaken. Nonetheless, with opportunistically available data sources, we are able to reproduce with reasonable accuracy the actual experiences for a diversity of alcohol-related acute outcomes in Fairfax County, Virginia. The novelty of the framework we have proposed lies in the ability to modify the conditional probabilities on the various paths through the directed graph and thus experimentally determine what social interventions are most likely to have a beneficial effect for reducing acute outcomes. Although this model is developed in terms of acute alcohol-related outcomes, there is an obvious applicability to expanding this framework to other settings such as drug-related behaviors, public health, and criminal activities.

## Acknowledgements

We would like to acknowledge the contributions of Dr. Rida Moustafa and our students, Peter Mburu and Walid Sharabati, who assisted us with the development of various figures and computations in this paper. Dr. Said's work is supported in part by Grant Number F32AA015876 from the National Institute on Alcohol Abuse and Alcoholism. The work of Dr. Wegman is supported in part by the U.S. Army Research Office under contract W911NF-04-1-0447. The content is solely the responsibility of the authors and does not necessarily represent the official views of the National Institute on Alcohol Abuse and Alcoholism or the National Institutes of Health.

## References

1. Carpenter C (2003) Seasonal variation in self-reports of recent alcohol consumption: racial and ethnic differences. J Stud Alcohol 64(3):415-418
2. Cho YI, Johnson TP, Fendrich M (2001) Monthly variations in self-reports of alcohol consumption. J Stud Alcohol 62(2):268-272
3. Ezatti M, Lopez A, Rodgers A, Vander Hoorn S, Murray C, and the Comparative Risk Assessment Collaborating Group (2002) Selected major risk factors and global and regional burden of disease. Lancet 360:1347-1360
4. Fitzgerald JL, Mulford HA (1984) Seasonal changes in alcohol consumption and related problems in Iowa, 1979-1980. J Stud Alcohol 45(4):363-368
5. Rehm J, Room R, Monteiro M, Gmel G, Rehn N, Sempos CT, Frick U, Jernigan D (2004) Alcohol. In Comparative Quantification of Health Risks: Global and Regional Burden of Disease Due to Selected Major Risk Factors. (Ezatti M, Lopez AD, Rodgers A, Murray CJL eds.). Geneva: WHO.
6. Said YH, Wegman EJ (2007) Quantitative assessments of alcohol-related outcomes. Chance 20(3):17-25
7. Wegman EJ, Said YH (2009) A Social Network Model of Alcohol Behaviors. This Volume

# HBML: A Language for Quantitative Behavioral Modeling in the Human Terrain

Nils F Sandell, Robert Savell, David Twardowski and George Cybenko

[first.last]@dartmouth.edu, Thayer School, Dartmouth College, Hanover, NH

**Abstract** Human and machine behavioral modeling and analysis are active areas of study in a variety of domains of current interest. Notions of what *behavior* means as well as how behaviors can be represented, compared, shared, integrated and analyzed are not well established and vary from domain to domain, even from researcher to researcher. Our current research suggests that a common framework for the systematic analysis of behaviors of people, networks, and engineered systems is both possible and much needed. The main contribution of this paper is the presentation of the Human Behavioral Modeling Language (HBML). We believe that the proposed schema and framework can support behavioral modeling and analysis by large-scale computational systems across a variety of domains.

## 1 Introduction

Quantitative human and machine behavioral modeling and analysis are becoming active areas of study in computer security [2, 13], insider threat detection [11], sensor network analysis [8, 1], finance [7, 12], simulation [5], and social network analysis [3] as well as other fields of great current interest. The basic concept of a "behavior" is intuitively easy to grasp, but the precise, quantitative notions of what behaviors are and how they can be represented, compared, shared, integrated and analyzed are not well established. They typically vary from domain to domain, even from researcher to researcher. In this paper, we propose definitions and a framework which we believe can accelerate progress on computational and algorithmic behavioral modeling and analysis.

Our belief that a common framework for discussing behaviors of people, networks and engineered systems is possible and is much needed has been informed by our previous research on process modeling and detection [2, 1, 12, 3]. Examples drawn from a familiar intelligence venue are presented as evidence to support this claim.

H. Liu et al. (eds.), *Social Computing and Behavioral Modeling*,
DOI: 10.1007/978-1-4419-0056-2_23, © Springer Science + Business Media, LLC 2009

## 2 Language Overview

The Human Behavioral Modeling Language (HBML) is distinguished from other behavioral modeling languages such as UML by being primarily a language for instantiating process models for profiling and tracking dynamic objects in the Human Terrain. Thus, the analysis of the represented objects and their dynamics is *data driven* rather than simulation oriented. The language incorporates generic process modeling and tracking techniques such as those forming the basis of the *Process Query System (PQS)*[4]. Experience in process tracking applications in diverse domains indicates that elements of a useful behavioral modeling system tend to fall into three broad generic categories:

- Environment: the rule system governing interactions among entities.
- Entities: the objects of interest- with their functional descriptions captured in one or more profiles.
- Behaviors: the space of potential processes in which an entity may be engaged.

The language is organized along two dimensions. The first dimension is scalar, with functional objects organized in multiple layers of aggregation from low level "atomic objects" to progressively larger scales of "aggregate objects". With properly defined dimension reductions at scale, aggregation may theoretically extend from individuals (atomic entities) through multiple scales of functional groupings, possibly to aggregate entities operating at the global scale. For example, a multi-scalar aggregation in the business domain might be defined in terms of a progression of discrete scales, such as: worker to workgroup to division to subsidiary to multi-national conglomerate. In addition, each entity is defined in terms of i) relevant stationary attributes, ii) the entity's locus in the human terrain, and iii) a functional description of potential behaviors in which the entity may be engaged. These behavioral descriptions define a second organizational dimension. Using a generic Shannon stochastic

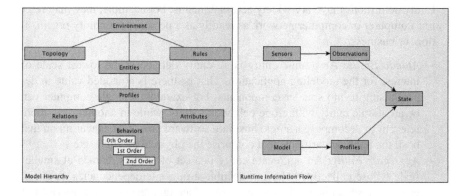

Fig. 1: HBML Object Hierarchy and Information Flow Diagram.

process framework, behaviors are described in terms of the 0th, 1st, and 2nd order descriptions of their process states and interactions.

These entities interact in an environment which encodes the topology and rule system determining the interactions of the objects. Examples of environmental systems include: i) communication network infrastructures, ii) electronic auction systems, iii) transportation and utility systems iv) political systems and v) social network structures.

## 3 Glossary of Terms and Concepts

In this section, we provide a more detailed description of the modeling language and its components which are necessary to achieve this goal. A diagram of the HBML hierarchy is given in Figure 1a, and an information flow diagram describing the estimation of entity state at runtime is given in Figure 1b. We begin by defining the rules of the system of interest via the environmental description.

- **Environment:** As described in the previous section, the environment is the "universe" through which entities navigate. Essentially, this corresponds to the problem domain and its mechanics. The environmental definition consists of a description of the environmental topology (when applicable) as well as an enumeration of the rules governing transactions or interactions among entities operating in the environment. If the environment encompasses a multi-scalar system, then one expects both inter- and intra-scalar rules to define the interactions among entities at appropriate levels of aggregation.

Operating within the bounds of the environment are some number of objects or entities which may be atomic or, alternatively, defined in terms of aggregates of entities at a smaller scale.

- **Entity:** Entities are objects in the environment capable of exhibiting behaviors. Entity definitions may vary widely across domains. For example, they may refer to a computer or computer network as readily as a person, community organization, or business.

  - *Atomic Entity:* An atomic entity is an entity defined at the lowest scale of interest for the modeling application. That is, there is no added value in decomposing it into smaller components. For example, a single computer can be an atomic entity within one behavioral study, while in another, it is more relevent to decompose a single host into software processes executing on that host, the hardware artifacts that comprise the physical machine, etc.
  - *Aggregate Entity:* An aggregate entity is a set of entities defined at smaller scale whose collective behaviors exhibit strong interdependencies. A social group in a social network is an example of an aggregate entity comprised of simpler atomic entities (people). A corporation is an example of an aggregate entity comprised of other aggregate entities (divisions of the corporation).

Characteristics of an entity are captured in one or more *profiles* which encode the entity's *attributes*, its *behavioral interface* and its *relationship* to other elements in the entity hierarchy. The behavioral interface is further developed according to Shannon's stochastic process framework in terms of ther 0th, 1st, and 2nd order descriptions of their process states and interactions.

- **Profile:** A profile is a description of an entity which is applicable to a particular context. Profiles contain information relevant to a specific behavioral modeling context- including behavioral descriptions, attributes and activities (described below) or distributions thereon. Multiple profiles may be assigned to a single entity-introducing a layer of abstraction into the entity definition which serves to simplify model encodings in many domains. For example, an entity object associated to a particular individual might be encoded in terms of three separate profiles- one for spatial navigation, one for computer use, and one for social interactions.

  - **Attribute:** Attributes are properties or meta data which describe an entity. They may be associated with an entity's existential status such as gender, age, height- and possibly include more transient properties such as mood. Attributes may also refer to values assigned by outside sources, such as a threat score. Attributes may take the form of a distribution over a set of values.
  - **Relations:** An entity's graphical and/or procedural interdependence with other entities is captured via definition of its relations. These are typically encoded as up, down, and side-links among entities. In familiar social examples, ancestor, descendant and friend or sibling relationships define a link structure (and thus a social network) upon the set of entities in the environment. However, entity relations may encode many other types of interdependence-as defined by the specific attributes associated with the edges.
  - **Behaviors:** Behaviors are the focus of HBML, and an entity's behavioral profile defines the range of procedures or activities available to an entity in a particular context. HBML employs a process taxonomy adapted from Shannon Information Theory [6], in the form of $n$th order behaviors as outlined below. Loosely speaking, a behavior consists of a set of activities together with a statistical characterization of those activities. Each behavior may be considered to be stationary within a certain time scale, or evolving according to some approximately known function.

The taxonomy of $n$th Order Behaviors is defined as follows:

- **0th Order Behaviors:** A 0th Order Behavioral Description is an enumeration of the processes or activities in which the entity may be engaged. One may think of the 0th Order Behavior as the entity's functional interface. For example, a 0th Order Behavior could be the list of activities a user may engage in on their computer, such as: {"visit website X", "read e-mail", "access file server", etc}. An apriori distribution may be defined over the space of 0th order behaviors, establishing the likelihood of selecting a particular behavior randomly from a profile in the absence of further contextual knowledge.

- **1st Order Behaviors:** A 1st Order Behavioral Description is an enumeration of the 0th Order activities in which the entity may be engaged, along with a posterior distribution describing the probability of engaging in the activity— conditioned on the time of day, spatial location or other environmental context. For example, when an individual is in their office in the morning, they check their e-mail, look at the morning news, or return to one of yesterday's tasks with some frequency.
- **2nd Order Behaviors:** A 2nd Order Behavioral Description is a probabilistic characterization of activities which is conditioned on other activities, rather than simply conditioned on environmental properties (as in a 1st order behavior). For example, behavioral models which characterize the probability that certain activities occur, conditioned on other activities having occurred previously, are 2nd Order behavioral models. Often, these models are *Markovian* in the sense that the probability distribution of activities depends on an entity's activities in the previous timestep. Models may also describe the temporal evolution of an entity's processes explicitly conditioned on other activities within the environment. For example, a 2nd order model provides the probability of a user printing a file after accessing the file server, or the probability that an employee attends a meeting given the fact that one of his workgroups is hosting one.

## 4 Anomaly Detection, Classification, and Prediction in HBML

In statistical behavioral modeling, the representation of behavior is essentially described as a probability distribution conditioned on past behavior and the appropriate contextual details:

$$p(x(t)|t, x(t-1), \ldots, x(0), \eta(t), \eta(t-1), \ldots, \eta(0)) \tag{1}$$

Here $x(t) \in X$ where $X$ corresponds to some set of labels and/or vectors that define and quantify the behavior of interest and $\eta(t)$ is a set of contextual information. Typically, assumptions are made to limit the number of arguments in this conditional probability. For example, assuming the behavior is time-homogenuous and 1st order Markovian, the distribution becomes $p(x(t)|x(t-1), \eta(t), \eta(t-1) \ldots, \eta(0))$ (a 2nd Order Model). It is also possible that a simpler contextual dependence is sufficient to capture our problem and we need only model the behavior as a function of the surrounding context, i.e. $p(x|\eta)$ (a 1st order model). Or perhaps even more simply, the apriori probabilities $p(x)$ (a 0th Order Model) will suffice.

We have found in our past experience with process modeling that this framework encompasses sufficient complexity to solve most, if not all, problems encountered in each of the three major tasks of human behavioral analysis: i) *Anomaly Detection*, ii) *Classification*, and iii) *Prediction*. In this section, we provide a brief overview of these three tasks and present supporting examples drawn from a familiar security domain.

## 4.1 Background: Terrorist Suspects in the Human Terrain

We demonstrate the flexibility of the HBML framework by examining an instance of each of the three types of tasks in the context of an intelligence example defined on the Human Terrain- that of the social, spatial, and functional analysis of cells of terrorists embedded in a social network.

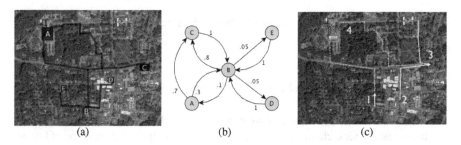

Fig. 2: (a) Geographical information with transit paths (b) Markov representation (c) Simple scenario

Consider a scenario in which groups of suspected terrorists and known associates are being surveilled in their movement throughout a city. Human observers stationed in the city, UAVs, and stationary surveillance equipment generate sensing reports of the locations of the individuals. Observations could be limited by line of sight occlusion (traveling through tunnels, buildings, etc) and equipment availability (a limited number of UAVs, agents). In order to aid in the analysis of the suspected terrorists' activities, we characterize their spatial and social behavior using HBML.

Figure 2a shows the labeled satellite imagery spanning the geographic area of our targets' regular daily activities. A single target of interest in the network has multiple frequent locations of interest, and a tendency to transition from those locations based on time of day. These locations are determined to be the target's place of work (A), residence (B), and 3 places for social gathering (C,D,E). The darkened paths indicate the usual paths taken between the locations of interest. Figures 3a and b represent the social network around the target, and the first order information about the location conditioned on time, respectively. The location distribution during the weekdays business hours has a very low entropy. Assuming the target's place of work is unaffiliated with criminal behavior, this target may be considered a criminal with a day job. Figure 3b shows a Markov chain governing the targets locational transitioning given that it is a weekday evening and 'Work Group 2' members are in Location C. There are a number of interesting problems that may be formulated, given this scenario. We start by describing the scenario in terms of the HBML representation framework.

- Atomic Objects - Persons of interest with known or suspected criminal ties.

185

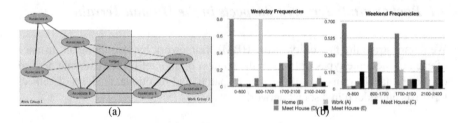

Fig. 3: (a) Social network surrounding target (b) Locational frequencies of target conditioned on time

- Aggregate Objects - Social networks with functional subgroups such as terrorist cells.
- Attributes - Name, background information, skills that may relate to criminal activity (electronics, chemicals, etc). Licensing or descriptive information of known vehicles used. Physical location or suspected location of a person.
- Relations - Membership in short-term social groups (i.e. meeting with other targets of interest). Location in the social network and functional groups, as well as strength-of-ties to other individuals. Ties between organizations.
- 0th Order Behavior - Locations a person is known to visit - residence, work, etc. Activities associated with groups such as meeting, synchronizing.
- 1st Order Behavior - Frequency an individual is at any location or a group meeting. Frequency of group activities, conditioned on time of day, day of week, etc.
- 2nd Order Behavior - Sets of Markov chains describing the frequency of transiting from location to location, conditioned on the locations and activities of known associates or functional groups.

### 4.1.1 Anomaly Detection

In *anomaly detection*, we develop a distribution over behaviors as suggested by Equation 1 and define it to be representative of 'normal' behavior. The goal of anomaly detection is to determine if a behavior set $x(t)$ is 'abnormal' in some statistically significant way. Typically, this is accomplished via some type of hypothesis testing. Anomaly detection can be thought of as a classification problem with one defined class (normal) and a null class defining abnormal behavior. Despite this, we consider the task to be semantically different from classification in that the only behaviors of interest are the abnormal ones (the ones falling in the null class).

In the context of our example, a deviation from "normal" behavior may indicate a change in the living conditions of the surveilled individuals. This change could be explained away by benign phenomena, such as relocation of living quarters or increased contact with new friends whom aren't suspected of any criminal activity. The abnormal behavior could also be indicative of the approach of an event of interest, such as a terrorist attack. In the days leading up to an attack, there may be an observable increase in meetings between various members of the group— as demonstrated by significant deviations from the historical frequency distributions of Figure 3b. In deploying an anomaly detection system, one might also consider joint statistics across the group, since it is possible that one member deviating from his routine is insufficiently abnormal, but a number of the group members changing their routine roughly simultaneously may be very anomalous and indicate an upcoming event of interest.

### 4.1.2 Classification

Consider an enumerated set of distinct profiles of behaviors associated with entities $1 \ldots \mathcal{N}$ and denote the behavioral 'state' of entity $i \in \mathcal{N}$ as $x_i(t)$. The classification problem is to assign the behavior of each entity $i$ to one or more of a finite set of classes $\mathcal{G}$ or perhaps a null class $g_0 \notin \mathcal{G}$. If the classes are known and determined *a priori*, then it is considered *classification*. If the classes are not defined *a priori*, then the task falls in the algorithmic domain known as *clustering*. In clustering, one simply needs to define a distance function over a probability distribution associated with the $x_i(t)$ values (generally, a marginal of Equation 1). To classify, each class $g \in \mathcal{G}$ requires a distribution $p_g(x(t)|\ldots)$ for comparison, where behaviors considered 'different enough' in some fashion from all classes is assigned to the null class. Many successful techniques exist, both for clustering (K-means, spectral, ...) and for classification (support vector machines (SVM's, decision trees, neural networks...).

There are a number of classification problems that are immediately recognizable in our example environment. First, it would be desirable to recognize functional subgroups and their members from the structure of the social network, the characterization of the relationships, and the attributes of the entities. In the social network fragment depicted in Figure 3a, work groups are recognized based on strengths of association and patterns of co-location. The target entity is seen frequently meeting with both members of Work Group 1 and 2. Associate A may be associated with members of Work Group 1, but rarely as a group and never at the group's typical meeting space. Once groups are determined, it would also be useful to identify roles of the individual members. For example, the targeted entity is the sole member of both Work Group 1 and Work Group 2. This could indicate a leader or go-between for the two groups - the determination of which role may be assisted by attributes of the individual.

### 4.1.3 Prediction

Statistical prediction of suspects' future behaviors is a vital capability for most security operations. The problem of prediction can be regarded as trying to estimate the behavioral quantity $x(t)$ at some time $t > t'$. In other words, we are interested in computing an estimate of $x(t)$ using some distribution:

$$p\left(x(t)|t,x(t'),x(t'-1),\ldots,x(0),\eta(t'),\eta(t'-1),\ldots,\eta(0)\right)$$

When $t > t'$, we term this *prediction*, and when $t \leq t'$, we term it *filtering*. The maximal likelihood solutions to these problems for 2nd Order processes are related to the Viterbi algorithm where the distribution is modeled by a Hidden Markov Model (HMM) and the Kalman filter when the distribution is modified by a nondeterministic linear dynamical system.

Given our example scenario, we may define the "cat and mouse" style pursuit of suspects by security and law enforcement officials. This task falls generally into the familiar predictor-corrector category of pursuit algorithms- however with the additional complexity associated with tracking a responsive human target. A principled behavioral modeling methodology can greatly enhance prediction and filtering for spatial tracking in this scenario. Synergistically, the locational information of the target tracking can aid in determining the general behavior of a target, or the behavior of a target can aid in refining the estimation of the target's location.

Our human targets consist of individuals and groups of terrorists. We use known behavioral patterns of targets to project destinations via selection and application of the generative 1st and 2nd order behavioral models. Surveillance priorities may then be assigned based on the value of the target, the perceived threat and the confidence in the projected destination. Additionally, as in [10], social network information with functional subgroups would be used to aid in destination projection. If a group of people, tightly associated, transit roughly simultaneously to a common location, a meeting can be inferred. Surveillance could be focused on a subset of the group, intending to determine the meeting location with smaller commitment of resources. Predictions would be updated given observed direction of travel. For the purposes of this description, we assume the targets are unaware of surveillance; however, reasoning over pursuit and evasion games could be used to augment the model.

For example, during a given day, stationary surveillance observes the target heading north from his residence (represented as track 1 in Figure 2c). The 2nd order model indicates the target is almost certainly heading to work - the dashed line indicates this inference. Mobile surveillance is directed towards surveillance of the more unpredictable associates G and E from the social network fragment depicted in Figure 3a. Surveillance tracks these individuals (tracks 2 and 3) to location C, a known meeting house for Work Group 2. A collection of this work group at location C is inferred as a meeting, and later on the target's 2nd order model conditioned on this meeting is fairly indicative that the target will head there as well (track 4). If a target is lost, 1st order frequency data may be used to prioritize the search to reacquire the target.

While the specific data set for this example is small, the global surveillance picture may be tracking several hundreds of individuals across a city with a fairly quick reaction time required. A predictive system can be useful for allocating limited sensing resources to maintain the most certain tracking picture possible.

# 5 Conclusion

In conclusion, the HBML framework promises to be a useful paradigm for encoding behavioral modeling tasks. We feel that it may serve as the basis for elucidating the structure which is common across diverse behavioral modeling domains. It is our hope that this presentation will begin a dialogue on the development of systematic methodology for analysis of behaviors of people, networks, and engineered systems.

## Acknowledgements

The authors wish to express their appreciation to Vincent Berk, Ian Gregorio-de Souza, James House, Alexy Khrabrov, John Murphy, and David Robinson for *helpful suggests and discussions* [9].

## References

1. V. Berk and G. Cybenko. Process query systems. *IEEE Computer*, pages 62–70, January 2007.
2. V. Berk and N. Fox. Process query systems for network security monitoring. In *Proceedings of the SPIE 0277-786X Defense and Security Symposium, Orlando, Florida*, March/April 2005.
3. W. Chung et al. Dynamics of process-based social networks. In *Sensors, and Command, Control, Communications, and Intelligence (C3I) Technologies for Homeland Security and Homeland Defense V*, Orlando, FL, April 2006.
4. G. Cybenko, V. H. Berk, V. Crespi, G. Jiang, and R. S. Gray. An overview of process query systems. In *Proc. SPIE 5403: Sensors, and Command, Control, Communications, and Intelligence*, pages 183–197, 2004.
5. S. Das. Intelligent market-making in artificial financial markets. PhD Thesis, Dpmt EECS, MIT, 2005.
6. R. W. Lucky. *Silicon Dreams*. St. Martin's Press, New York, 1989.
7. A. Madhavan. Market microstructure: A survey. *Journal of Financial Markets*, 3:205–258, 2000.
8. S. Oh, L. Schenato, P. Chen, and S. Sastry. A scalable real-time multiple-target tracking algorithm for sensor networks. University of California, Berkeley, Technical Report UCB//ERL M05/9, Feb. 2005.
9. A. Putt. *Putt's Law and the Successful Technocrat: How to Win in the Information Age*. Wiley-IEEE Press, 2006.
10. R. Savell and G. Cybenko. Mining for social processes in intelligence data streams. *Social Computing*, Jan 2008.
11. S.Stolfo et al. *Insider Attack and Cyber Security: Beyond the Hacker*. Springer, New York, March, 2008.
12. D. Twardowski, R. Savell, and G. Cybenko. Process learning of network interactions in market microstructures. submitted for publication, preprint available, 2008.
13. B. Warrender and S. Forrest. Detecting intrusion using system calls: alternative data models. In *1999 IEEE Symp. on Security and Privacy*, pages 133–145, Oakland, CA, 1999.

# VIM: A Platform for Violent Intent Modeling

Antonio Sanfilippo, Jack Schryver, Paul Whitney, Elsa Augustenborg, Gary Danielson, Sandy Thompson

{antonio.sanfilippo, paul.whitney, elsa.augustenborg, gary.danielson, sandy. thompson}@pnl.gov
Pacific Northwest National Laboratory, 902 Battelle Blvd., Richland, WA

schryverjc@ornl.gov
Oak Ridge National Laboratory, 1 Bethel Valley Road, Oak Ridge, TN

**Abstract** Radical and contentious activism may or may not evolve into violent behavior depending on contextual factors related to social, political, cultural and infrastructural conditions. Significant theoretical advances have been made in understanding these contextual factors and the import of their interrelations. However, there has been relatively little progress in the development of processes and capabilities that leverage such theoretical advances to automate the anticipatory analysis of violent intent. In this paper, we describe a framework that implements such processes and capabilities, and discuss the implications of using the resulting system to assess the emergence of radicalization leading to violence.

## 1 Introduction

The history of social movements is ripe with examples of contentious/radical groups and organizations that share the same ideology and yet through time develop opposite practices towards the use of violence. For example, compare al-Gama'a al-Islamiyya with the Muslim Brotherhood in Egypt during the 1990s [1] or the Red Brigades with extra parliamentary groups such as the Italian Marxist-Leninist Party in Italy during the 1970s [2]. In both cases, we have radical ideologies, one aiming at the establishment of a theocratic state ruled by Shariah law (al-Gama'a al-Islamiyya and the Muslim Brotherhood) and the other targeting the establishment of a totalitarian state ruled by a proletarian dictatorship. Yet, while the Muslim Brotherhood and Italian Marxist-Leninist Party have not pursued terrorism as a means to attain their political goals, al-Gama'a al-Islamiyya and the Red Brigades chose the opposite mode of action culminating in tragic events such as the 1992 Luxor massacre in Egypt where 58 foreign tourists and 4 Egyptians were killed, and the murder of former Italian prime minister Aldo Moro in 1978.

H. Liu et al. (eds.), *Social Computing and Behavioral Modeling*,
DOI: 10.1007/978-1-4419-0056-2_24, © Springer Science + Business Media, LLC 2009

Substantial progress has been made across a variety of disciplines including sociology, political science and psychology to explain the emergence of violent behavior. The theoretical advances in research areas such as social movement theory [3], leadership trait analysis [4], conceptual/integrative complexity [5], theories of social identity such as group schism/entitativity [6], and frame analysis [7] now provide valuable insights that can be used to develop analytical tools and models to explain, anticipate and counter violent intent. Current approaches tend to focus either on the collection and analysis of relevant evidence [8], or the development of predictive models based on expert knowledge [9, 10]. However, little effort has been devoted so far to the development of a software platform with associated workflows which combines expert knowledge encapsulation, evidence extraction and marshaling, and predictive modeling into a cohesive analytical discourse. The availability of such computational tools can greatly enhance the ability of humans to perform strategic/anticipatory intelligence analysis. Because of memory and focus attention limitations on human cognition [11, 12], most people are unable to satisfy the essential requirements for unbiased judgment by retaining several hypotheses and relevant supporting evidence in working memory. Moreover, supporting evidence needs to be distilled from potentially huge data repositories. Such a task requires extravagant expenditure of human resources without the help of machine-aided data analysis processes, and may thus not result in timely actions. The goal of this paper is to describe the development and functionality of a software platform that helps overcome some of the main human limitations in the anticipatory analysis task, with specific reference to violent intent modeling

We begin with a review of analytical tools and models which have been recently developed to understand and anticipate violent intent. We then describe a new approach which provides a more comprehensive platform through the integration of capabilities in expert knowledge encapsulation, evidence extraction and analysis, and predictive modeling within a service-oriented architecture. We conclude by commenting on the value of this novel approach for intelligence analysis with specific reference to violent intent modeling.

# 2 Background

Significant progress has been made in the collection, marshaling and dissemination of evidence relevant to the study or terrorism. For example, the *National Consortium for the Study of Terrorism and Response to Terrorism*(START) [8] has recently made available three databases that organize and streamline access to existing data resources on terrorism: The *Global Terrorism Database,* which provides information on terrorist events worldwide since 1970; the *Terrorism & Preparedness Data Resource Center*, which archives and distributes data about domestic and international terrorism, responses to terror and natural disasters, and attitudes towards terrorism and the response to terror; and the *Terrorist Organiza-*

*tion Profiles*, which provide access to data on terrorist groups and their key leaders.

On the modeling side, advances have been made both in agent-based and probabilistic evidentiary reasoning approaches. For example, Chaturvedi et al. [10] present an agent-based modeling approach to the analysis of dynamic interrelationships among grievances, level of resources, and organizational capacity to mobilize members toward insurgency using insights from resource mobilization theory [13]. This approach requires the calibration of agents with published data relevant to resource mobilization theory. Sticha et al. [9] describe an application of Bayesian nets that enables the analyst to perform anticipatory reasoning on a subject's decision-making process using as indicators personality factors derived from Leadership Trait Analysis [4] and the Neuroticism-Extroversion-Openness (NEO) Inventory [14]. Indicators are instantiated with evidence drawn from (a) running the NEO test [14] on input forms compiled by subject matter experts, and (b) processing first-person verbalizations with Profiler+ [15] to obtain leadership personality profiles.

These approaches and related works present novel ways of addressing the study of terrorism. However, each analytical platform addresses specific aspects of the task at hand. For example, the START efforts focus on data collection, analysis and dissemination, and the primary focus of the agent-based and Bayesian nets approaches developed by Chaturvedi et al. and Sticha et al. is modeling. While sharing many similarities with each of these approaches, the framework described in this paper strives at providing a more encompassing solution in which activities such as data collection, expert knowledge encapsulation, evidence extraction and analysis, and modeling are integrated into a cohesive workflow.

## 3 The VIM Approach

A recent effort to provide an integrated suite of analytic methods for anticipatory/strategic reasoning is the Violent Intent Modeling (VIM) project. The goal of the VIM project is to help analysts assess the threat of violence posed by contentious radical groups and organizations through the development of an integrated analytical framework that

- Leverages the expertise of social scientists to model ideological, organizational, and contextual factors associated with violent intent
- Uses probabilistic evidentiary reasoning to assess violent intent hypotheses
- Combines data collection and content extraction and analysis methods to identify signatures of violent intent in data sets
- Uses knowledge representation and presentation methods to facilitate the modeling task by encapsulating and providing access to the expert knowledge and evidence gathered
- Facilitates ubiquitous user access to components and workflows through a web-enabled interface.

Consequently, VIM includes the following components.

- A social science guidance map that provides access to theoretical approaches and related case histories relevant to the analysis of violent intent
- A content extraction and analysis component that facilitates the extraction and measurement of relevant evidence from document repositories to inform models of violent intent
- A predictive modeling component based on Bayesian nets interpretation of Heuer's Analysis of Competing Hypothesis [12] capable of leveraging social science theories and building models of violent intent informed by evidence
- A web enabled software integration platform that combines the relevant modeling and knowledge/evidence capture/access capabilities and provides tools for content management and collaborative work.

Progress along these lines is anticipated to provide non-linear advances in how analysts can leverage technology and analytic science in their work.

## 4 Social Science Guidance Map

The objectives of the Social Science Guidance Map (SSGM) are to support analysts in identifying relevant social science theories and concepts related to the analytical task, and to assist in generating alternative hypotheses and perspectives about an adversary's likely behavior. The SSGM presents summaries of social and behavioral science topics relevant to understanding general principles of group behavior, as well as research related more specifically to estimating the likelihood that a group will adopt violence as a strategy to achieve its objectives. There are summaries of 34 social science topics currently implemented in the SSGM. These summaries enable users who may only have a basic, college-level knowledge of the social sciences to learn about and utilize social and behavioral factors relevant to violent intent modeling within the VIM system.

The SSGM goes beyond encyclopedic sources of knowledge such as Wikipedia in three important ways: (1) by providing not just summaries of social science topics, but practical assistance on the application of social science to intelligence analysis; (2) by displaying concept maps or diagrams of social science knowledge that will allow users to easily navigate through and comprehend a massive conceptual space without becoming cognitively lost; and (3) by compiling observable and collectable indicators of important social science constructs.

The SSGM is organized around the principles of hypertext [16] and concept mapping, which offer a powerful method of representing the cognitive structure of a knowledge domain in the form of a diagram composed of labeled nodes (concepts) and links (relationships). This structure is sometimes referred to as a cognitive scaffold [17] and is used as a training device that allows learners to glimpse "under the hood" in order to reveal an expert's organization of the material. Concept maps are intuitive graphic displays that reveal how to "connect the dots" in a massively sized conceptual space [18, 19]. The SSGM contains a collection of topical wiki pages that integrate textual material in a nonlinear hyperlinked environment with multiple entry points to facilitate browsing.

The VIM system contains a recommender interface that matches features of analyst problems to social science topics and models. The recommender algorithm uses a large matrix which relates social science concepts and models to statements expressed in plain English (C-S matrix). The C-S matrix stores quantitative weights that reflect the relevance of each statement to a specific social science concept or model. Users respond to statements by indicating their degree of agreement on a Likert-type scale from 1 to 7 (see "Guided Exploration" in Figure 5). The relevance of a social science concept or model to a statement is calculated using a linear combination of weighted user responses to all statements by referencing values in the C-S matrix. A rank-ordered list of recommendations is then presented to the user. Each social science concept in the rank-ordered list is directly hyperlinked to the wiki page for that concept. This feature allows users to navigate directly to the most relevant social science topics.

As shown in Figure 1, the main levels represented in the SSGM are individual, group, state-society as well as interactions between levels and can be easily explored in details in a drill-down fashion. The SSGM encourages a holistic approach to analysis of social problems, which is more conducive to an analyst perspective than a single-discipline level of analysis and more appropriate for topics that cross-cut the social sciences. For example, framing has been studied extensively in sociology within the context of social movement theory, mass communication, and cognitive psychology with reference to prospect theory. In social movement theory, framing is described as a technique used by a social movement entrepreneur to mobilize a target population. Communication theory focuses more upon how mass media frame thematic content. In prospect theory, framing refers to a proc-

**Figure 1.** Users can retrieve a concept map for further drill-down by clicking an icon in the main SSGM page.

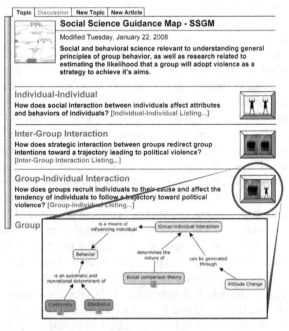

ess by which different encodings of a decision problem modify utilities–e.g. a fixed frame of reference is typically used by decision makers to evaluate utilities of outcomes. These interpretations of framing are all represented in the SSGM, located in different levels of the concept map hierarchy.

The SSGM captures the tendency by many social science theories to revolve about central constructs that provide a cohesive framework for causal explanation, e.g. the wiki page on group schism invokes the "group cohesion" construct. The SSGM attempts to associate such central constructs with indicators, i.e., observable, collectable or measurable metrics to serve as proxies for more abstract concepts. For example, one of the indicators for group cohesion in the SSGM is specified as "membership turnover rate"–a quantifiable index that can be estimated by integrating reports from open sources. Indicators can be linked to models so that users can more easily access conceptual background with consequent gains in model and inferential transparency.

## 5 Content Extraction and Analysis

The Content Extraction and Analysis component (CEA) provides a web-enabled, semantically-driven and visually interactive search environment, as shown in Figure 2. CEA combines an information extraction pipeline with the Alfresco content management environment and the SQL Server database (see section 7). The information extraction pipeline consists of a variety of natural language components which perform event detection with added specifications of named entities (people, organizations, locations) which act as event participants, word domains, and dates [20]. Currently, event detection covers some 4000 verbs, organized into a 3-tired hierarchy with a top layer of 31 event types branching out into some 120

**Figure 2.** CEA interface in VIM.

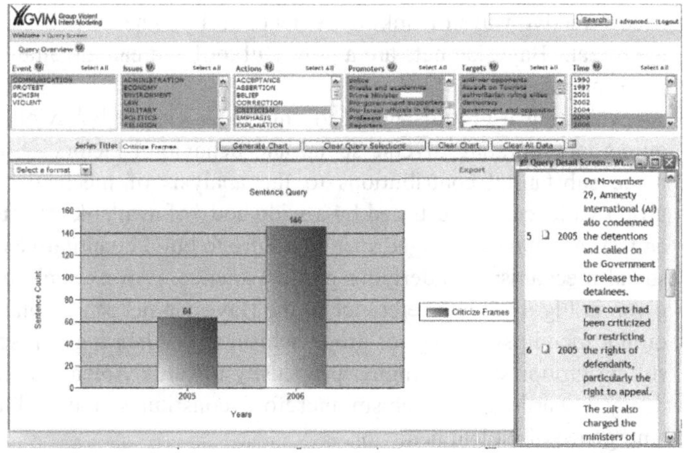

event subtypes. This hierarchy can be easily tailored to specific analytical needs, such as the identification of group schism events which would inform the model described in section 6. Nearly ¼ of the event hierarchy is devoted to the automation of frame analysis to help identify the strategies that communication sources adopt to influence their target audiences [20, 21]. For example, such an analysis may tell us about the distribution of contentious vs. negotiation framing strategies that a group exhibits from year to year, as indicated in Table 1. CEA results can be used to inform the modeling task by providing quantified evidence relative to specific indicators, e.g. textual evidence of events describing collective action frames with scores indicating the relative strength of the evidence observed.

Table 1. Frame distributions for Group-X in years X and X+1 (adapted from [21]).

|  | Contentious Frames | Negotiation Frames | z-score |
|---|---|---|---|
| Year X | 24% (43/177) | 14% (26/177) | 2.28 |
| Year X+1 | 35% (77/217) | 10% (23/217) | 6.16 |
| z-score | 2.40 | 1.22 | |

## 6 Bayesian Analysis of Competing Hypotheses

A Bayesian net (BN) is a graphical representation of causal relationships among variables. BN's are compact methods of representing causal interactions in complex environments, where uncertainty is a predominant feature. BN's are especially useful in diagnostic systems that determine the likelihood of observation-based hypotheses. BN's are used in many analytical applications ranging from tools for medical diagnosis, e.g. to relate probabilistically the occurrence of a disease to the diagnostic evidence (symptoms), to decision support for military combat, e.g. Weapons of Mass Destruction-Terrorism (WMD-T) rare event estimation. In such applications, analysts are presented with a variety of evidence points relevant to a threat (e.g. occurrence of an infectious disease or a WMD-T event) and need effective ways of linking evidence to hypotheses to assess the likelihood of the threat. Bayesian nets are a powerful tool and can aid analysts in making such assessments.

Evaluating a group's intent to engage in violent behavior is difficult and requires incorporating diverse social and behavioral factors. Social sciences have made substantial contributions to the analysis of mechanisms associated with group violence, as illustrated by a wide body of available literature (see sections 1 and 2). We have leveraged such literature to build quantitative models that encode social mechanisms underlying group violence. An example of such an effort is given in Figure 3 with reference to the Bayesian net representation of Sani's work on the social psychology of group schism [6]. When a group splits, one of the resulting groups shows a higher propensity towards violent behavior [22]. The tendency towards group schism therefore constitutes a useful diagnostic for the emergence of violent behavior.

Building Bayesian networks that represent reliable social and behavioral constructs requires identifying key factors from within social science theories and establishing relationships among such factors so that the Bayesian network model built reflects the insights of the social science theories used as sources. For instance, the model shown in Figure 3 was developed by distilling key factors from Sani's theory of group schism to establish the nodes within the model and identifying relationships across the factors in Sani's theory to link the nodes in the model. For example, the *voice* indicator conceptualizes the extent to which a faction within a group has effective means of expressing its position. A low value for *voice* signals the increasing likelihood that the group may splinter. Similarly, a higher value for the *intergroup differential or conflict* indicator can lead to reduction in group cohesion with consequent weakening of the group identity leading to higher likelihood that schism intentions may proliferate.

**Figure 3.** Bayesian net representation of model assessing likelihood of a group split. The nodes represent the critical concepts identified in the literature. In this model, each concept has two state values (Low/High, or Present/Absent). The numeric values are the marginal probabilities for the state values. The model propagates probability values through the network so that if a value changes (based on evidence or opinion) in one part of the model, corresponding changes in probability values are observed in the other parts of the model.

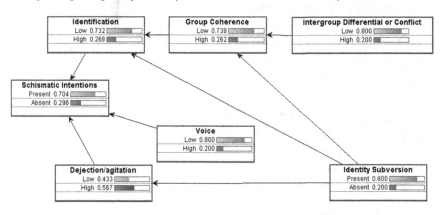

By quantifying indicators of violent intent that social science literature only qualitatively describes, Bayesian nets help analysts explain their analytic conclusions and reproduce them. Bayesian nets encourage analytic objectivity by explicitly reminding analysts of the larger set of relevant indicators.

# 7 The VIM Software Platform

The VIM software platform offers a framework for the integration of social science theories and models with associated content extraction and analysis processes, a sophisticated data repository, and a set of text analytics services.

The data repository is composed of a customized version of the Alfresco content management system and a relational database (currently SQL Server). All VIM data is stored in the data repository. The set of text analytics services was built using the technologies and pipeline specific to the CEA component described in section 5. Location, organization and person names are extracted from each document or document fragment collected and entered into the repository. This information is then readily available, via web services, to all components in the VIM system, as shown in Figure 4. A synonymy service, based on WordNet, is incorporated in the VIM repository to help expand search terms in their context. The VIM content repository extends Alfresco categories to allow analysts to use the social science taxonomies provided by the VIM system or create their own taxonomies.

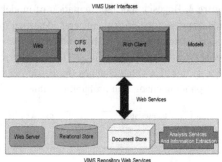

**Figure 4.** VIMS system architecture.

All VIM repository services are made available via web services. In this way, VIM software products can access the full functionality of the VIM repository regardless of the programming language or the platform used. This facilitates the creation of interfaces tailored to the client needs. The VIM system uses both web-based and rich client interfaces to accommodate its components.

The web interface shown in Figure 5 provides the basic tool for accessing expert knowledge, evidence, and model data. It allows the user to create tasks (basic units of analysis), search for evidence, and perform content analysis (see sections 4, 5 and 6 for examples). Finally, the recommender interface matches analysts' questions to social science concepts and models (see "Guided Exploration" in Figure 5 and section 4).

**Figure 5.** VIMS web interface.

The rich client interface was necessary because the modeling component described in section 6 is implemented as a rich client. Efforts are currently underway for a web-based conversion of the modeling component to enable full integration with the VIM web interface.

When an analyst creates a task, a set of template folders are automatically generated in the VIM Content Repository. The analyst selects the most appropriate models/theories for the task by either selecting from a list or by answering a set of questions in the recommender interface to receive suggestions about the most appropriate theories and models. The user may then launch the selected models or review the selected theories. The collection and analysis of evidence to be used in the models is enabled by the text analytics services and the CEA capabilities described in section 5.


# 8 Conclusions

If progress is to be made in the procactive assessment of violent intent, a software platform  has to become available that integrates mathematical modeling tools with workflows and capabilities able to leverage theoretical insights from the social sciences and facilitate the extraction and analysis of evidence from data streams. We have shown that such a platform can be developed by using a Wiki-based expert knowledge management approach and content extraction and analysis processes to inform a probabilistic evidentiary component within a service-oriented architecture that leverages both web-based and rich client interfaces. The resulting system offers a multidisciplinary approach to the study of violent intent deeply rooted in the relevant social and behavioral sciences that gives equal attention to mathematical modeling, the acquisition of knowledge inputs and the facilitation of cognitive access for users.


# References

1.  Al-Sayyid, M. K. (2003) *The other Face of the Islamist Movement.* Carnegie  Paper No. 33, January 2003. Available at http://www.carnegieendowment.org.
2.  Bocca, G, (1978) *Il terrorismo italiano*, Rizzoli, Milano.
3.  McAdam, D., S. Tarrow,  C.Tilly (2001) *Dynamics of Contention.* Cambridge University Press, New York, NY.
4.  Hermann, M. G. (2003) Assessing leadership style: Trait analysis. In Jerrold M. Post (Ed.) *The Psychological Assessment of Political Leaders: With Profiles of Saddam Hussein and Bill Clinton*, University of Michigan Press, Ann Arbor.
5.  Suedfeld, P., Tetlock, P., & Streufert, S. (1992). Conceptual/integrative complexity. In C.P. Smith (Ed.) *Motivation and Personality: Handbook of thematic content analysis.* Cambridge, England: Cambridge University Press.
6.  Sani, F. (2005). When subgroups secede: Extending and refining the social psychological model of schisms in groups. *Personality and Social Psychology Bulletin*, 31:1074-1086.
7.  Benford, D. and R. Snow (2000) Framing Processes and Social Movements: An Overview and Assessment. *Annual Review of Sociology*, Vol. 26, pp. 611-639.

8.  START: The National Consortium for the Study of Terrorism and Responses to Terror. University of Maryland. http://www.start.umd.edu.

9.  Sticha, P., Buede, D. and Rees, R. (2005) APOLLO: An Analytical Tool for Predicting a Subject's Decision-making. *Proceedings of the 2005 International Conference on Intelligence Analysis*, McLean, VA.

10. Chaturvedi, A., Purdue University, D. Dolk, R. Chaturvedi, M. Mulpuri, D. Lengacher, S. Mellema, P. Poddar, C. Foong,and B. Armstrong, (2005) Understanding Insurgency by Using Agent-based Computational Experimentation: Case Study of Indonesia. *Proceedings of the Agent 2005 Conference on Generative Social Processes, Models and Mechanisms,* Chicago, IL, pp. 781-799.

11. Miller, G. (1956) The Magical Number Seven, Plus or Minus Two: Some Limits on Our Capacity for Processing Information. *The Psychological Review*, 63: 81-9.

12. Heuer, R.J. Jr. (1999) *Psychology of Intelligence Analysis*. Center for the Study of Intelligence, Central Intelligence Agency, Washington, DC.

13. McCarthy, J. D. and Mayer N. Zald (2001) *The Enduring Vitality of the Resource Mobilization Theory of Social Movements* in Jonathan H. Turner (ed.), *Handbook of Sociological Theory*, pp.535-65.

14. Costa, P.T. & McCrae, R.R. (1985) *The NEO Personality Inventory manual*. Odessa, FL: Psychological Assessment Resources.

15. Young, M.D. (2001) Building worldview(s) with Profiler+. In M.D. West (ed.), *Applications of computer content analysis*. Westport, CT: Ablex.

16. Shapiro, A. and Niederhauser, D. (2004). Learning from hypertext: research issues and findings. In D. H. Jonassen (Ed.), *Handbook of Research for Educational Communications and Technology,* 2nd Ed., pp. 605-620. Mahwah, NJ: Erlbaum.

17. O'Donnell, A. M., Dansereau, D. F. And Hall, R. H. (2002) Knowledge maps as scaffolds for cognitive processing. *Educational Psychology Review*, 14(1), 71-86.

18. Nielsen, J. (1990). The art of navigating through hypertext. *Communications of the ACM*, 33(3): 297-310.

19. Tergan, S. O. (2004). Concept maps for managing individual knowledge. *Proceedings of the First Joint Meeting of the EARLI SIGS*, pp. 229-238.

20. Sanfilippo, A., A.J. Cowell, S. Tratz, A. Boek, A.K. Cowell, C Posse, and L. Pouchard (2007) Content Analysis for Proactive Intelligence: Marshaling Frame Evidence. *Proceeding of the AAAI Conference*. Vancouver, BC, Canada, July 22–26, 2007.

21. Sanfilippo, A., L. Franklin, S. Tratz, G. Danielson, N. Mileson, R. Riensche, and L. McGrath (2008) Automating Frame Analysis. In H. Liu, J. Salerno, and M.Young (eds.), *Social Computing, Behavioral Modeling, and Prediction*, pp. 239-248. Springer, NY.

22. McCauley, C., S. Moskalenko (2008) Mechanisms of Political Radicalization: Pathways Toward Terrorism. *Terrorism and Political Violence*, Volume 20, Issue 3 July 2008, pages 415 – 433.

# Punishment, Rational Expectations, and Relative Payoffs in a Networked Prisoners Dilemma

Shade T. Shutters[†]

[†] shade.shutters@asu.edu, Arizona State University, Tempe, Arizona

**Abstract** Experimental economics has consistently revealed human behavior at odds with theoretical expectations of rational agents. This is especially true in laboratory games with costly punishment where humans routinely pay to punish others for selfish behavior even though the punisher receives no benefit in return. This phenomenon occurs even when interactions are anonymous and the punisher will never interact with the punishee again. However, costly punishment may not be inconsistent with Darwinian notions of relative fitness. This paper presents exploratory work aimed at a reconciliation between economic and biological expectations of behavior. Agent-based modelling is used to simulate networked populations whose members play the prisoners dilemma while having the ability to altruistically punish one another. Results show that behavior evolving in structured populations does not conform to economic expectations of evolution driven by absolute payoff maximization. Instead results better match behavior expected from a biological perspective in which evolution is driven by relative payoff maximization. Results further suggest that subtle effects of network structure must be considered in theories addressing individual economic behavior.

## 1 Introduction

The mechanism of altruistic punishment is a leading candidate for explaining the evolution of cooperation. Altruistic punishment occurs when an individual incurs a cost to punish another but receives no material benefit in return [1-3]. It has been shown to induce cooperative behavior in numerous studies [3-6]. However, controversy regarding altruistic punishment lingers because, while it may explain many instances of cooperative behavior, the mechanism itself is seemingly irrational. Why should an individual expend fitness or wealth to punish someone with whom they will never again interact when they receive no benefit from doing so? Yet despite the economic prediction that a rational agent will not pay to punish others, humans repeatedly do so in laboratory experiments [7-9].

H. Liu et al. (eds.), *Social Computing and Behavioral Modeling*,
DOI: 10.1007/978-1-4419-0056-2_25, © Springer Science + Business Media, LLC 2009

This economic expectation is based on the widely-held premise that agents with independent preferences act to maximize their absolute payoffs[1]. Previously a framework was presented for expectations when the evolution of punishment behavior is driven by relative payoffs instead of absolute payoffs [11]. In this study, computer simulations were conducted to test this framework and to determine whether agents evolved behavior reflective of absolute payoff maximization or relative payoff maximization.

## 2 The Simulation

In a series of agent-based computer simulations, social networks comprised of regular lattices were populated with agents that played the continuous prisoners dilemma against one another. After each game, a 3rd party observer had the opportunity to pay to punish agents in the game if a player's contribution was deemed too low. Each component of an agent's strategy – how much to contribute in a game, when to punish, and how much to spend on punishment – were all independently evolving attributes of an agent.

### 2.1 The Continuous Prisoners Dilemma

The prisoners dilemma is a commonly used framework for studying social dilemmas and consists of a two-player game with payoffs structured so that social goals and individual goals are at odds. Total social welfare is maximized when both players cooperate but a player's individual payoff is always maximized by cheating.

In the classic prisoners dilemma players are limited to two choices - cooperate or defect. In this study agents select a level of cooperation $x$ at any point on a standardized continuum between full defection ($x = 0$) and full cooperation ($x = 1$). This is known as the continuous prisoners dilemma (CPD) and presents an arguably more realistic picture of the complexity of true social dilemmas [12]. In a game between agents $i$ and $j$, $i$'s payoff $p$ is

(1)     $p_i = 1 - x_i + r(x_i + x_j)/2; \quad x \in [0,1] , \ r \in (1,2)$

---

[1] Due to size limitations this paper does not review a large body of economics literature relevant to the current topic. Interested readers are encouraged to seek scholarly works dealing with relative preferences, other-regarding preferences, relative utility, conspicuous consumption, and positional goods, among others. See [10] for a lengthy introduction to the topic.

where $r$ represents the synergistic effect of cooperating. In this study $r = 1.5$ in all cases. Total social welfare $p_i + p_j$ is maximized when $x_i = x_j = 1$, yet $i$'s payoff is maximized when $x_i = 0$ regardless of the contribution made by $j$.

## 2.2 Altruistic Punishment

The introduction of punishment to the CPD adds a $3^{rd}$ party to the game – the observer. The decision of whether to punish or not is controlled by an attribute of the observer, the observer's punishment threshold, which is compared to a player's contribution in an observed game. If the player's contribution falls below the observer's threshold, the observer punishes the player. The amount the observer spends to punish $c$ is also an agent attribute but is only invoked when the agent acts as a $3^{rd}$ party observer of a neighboring game.

Letting $c$ = the cost to the observer to punish, the amount deducted from the punished player is $cM$, where $M$ is the punishment multiplier, a simulation parameter controlling the relative strength of punishment.

In agreement with many recent experiments, the introduction of punishment in structured populations led to cooperative behavior. However, the focus of this study is not on the cooperative outcomes of the prisoners dilemma but on the mechanism of altruistic punishment that induces those outcomes.

## 2.3 Social Structure

Testing predictions of the relative payoff model required simulations in which agents had different numbers of neighbors. For this purpose social networks were used to structure populations so that interactions, both game play and observation, were restricted to a fixed number of immediate neighbors. To control for confounding effects due to variance in the number of neighbors among agents, networks were limited to regular lattice structures (Table 1) so that variance in number of neighbors = 0 in all simulations.

**Table 1.** Lattice networks used in this study.

| Network Type | Description of Neighbors | Number of Neighbors |
|---|---|---|
| Ring | Left, right | 2 |
| Von Neumann | Left, right, up, down | 4 |
| Hexagonal | Left, right, diagonals | 6 |
| Moore | Left, right, up, down, diagonals | 8 |
| Complete | All other agents | $N$-1 |

In addition to regular lattices, simulations were run on a complete network. In a complete network every agent is linked to every other agent and, like a regular lattice, has no variance in number of neighbors per agent. A complete network represents the social network version of a homogeneous well-mixed system.

## 2.4 The Simulation Algorithm

A single simulation initiates with the creation of the appropriate lattice network. At each of 400 nodes is placed an agent $i$ consisting of strategy $(x_i, l_i, c_i)$ where $x_i$ = the contribution $i$ makes to the public good when playing against $j$, $l_i$ = the offer limit below which $i$ will punish another agent in a game being observed by $i$, and $c_i$ = the cost that $i$ incurs to punish the observed agent when the observed agent's offer is too low. In other words $l_i$ determines if agent $i$ will inflict punishment and $c_i$ determines how much punishment agent $i$ will inflict. Each of the three strategy components holds a value on the continuous interval $[0,1]$ and is generated randomly from a uniform distribution at the beginning of a simulation. To control for other factors that might contribute to the maintenance of cooperation, such as interaction history or reputation, agents have no recognition of or memory of other agents.

The simulation proceeds for 10,000 generations, each consisting of a game play routine, including observation & punishment, and a reproduction routine. During a single CPD game an agent $i$ initiates a game by randomly selecting $j$ from its neighborhood. Both agents are then endowed with one arbitrary unit from which they contribute $x_i$ and $x_j$ respectively to a public good. Players' choices are made simultaneously without knowledge of the others' contribution. $i$ then randomly selects a second neighbor $k$, who is tasked with observing $i$'s contribution. If $k$ judges the contribution to be too low, $k$ pays to punish $i$. After each game, running payoffs for $i$, $j$, and $k$ are calculated as shown in Table 2. Each agent initiates 3 games during a single generation and all games in a generation are played simultaneously.

**Table 2.** Payoffs $p$ for a CPD game between $i$ and $j$ with 3rd party punishment of $i$ by $k$, where $r$ is the synergistic effect of cooperation and $M$ is the punishment multiplier.

| | $x_i \geq l_k$ | $x_i < l_k$ |
|---|---|---|
| $k$ punishes $i$? | no | yes |
| $p_i$ | $1 - x_i + r(x_i + x_j)/2$ | $1 - x_i + r(x_i + x_j)/2 - c_k M$ |
| $p_j$ | $1 - x_j + r(x_i + x_j)/2$ | $1 - x_j + r(x_i + x_j)/2$ |
| $p_k$ | $0$ | $- c_k$ |

Following game play the reproduction routine runs in which each agent $i$ randomly selects a neighbor $j$ with which to compare payoffs for the generation. If $p_i > p_j$,

$i$'s strategy remains at $i$'s node in the next generation. However, if $p_i < p_j$, $j$'s strategy is copied onto $i$'s node for the next generation. In the event that $p_i = p_j$, a coin toss determines which strategy prevails. As strategies are copied to the next generation each strategy component of every agent is subject to mutation with a probability $m = 0.10$. If a component is selected for mutation, Gaussian noise is added to the component with mean = 0 and std. dev. = 0.01. Should mutation drive a component's value outside [0,1], the value is adjusted to the closer boundary value.

## 3 The Model of Relative Payoff Maximization

Because evolution is driven by relative fitness as opposed to absolute fitness, it is important to consider the possibility that evolution of economic behavior is driven by relative payoffs. In other words, letting $p$ = the payoff of agent $i$ and $P$ = the sum payoffs of the $n$ neighbors of $i$, we should compare evolved behavior when $i$'s survival is driven by maximization of $p$ (absolute payoff) versus maximization of $p/P$ (relative payoff). At the risk of scandalous oversimplification, we can think of these two approaches as the economic and biological viewpoints respectively.

Letting time step 0 = a point in time prior to an act of punishment and time step 1 = a point after punishment, the following describe the effects of a single act of punishment:

(2a)     $p_1 = p_0 - c$

(2b)     $P_1 = P_0 - cM.$

For an act of punishment to be evolutionarily beneficial it should lead to an increase in $i$'s relative payoff such that the following is true:

(3)     $\dfrac{p_1}{P_1} > \dfrac{p_0}{P_0}$     or     $\dfrac{p_0 - c}{P_0 - cM} > \dfrac{p_0}{P_0}.$

When simplified and restated [11], this model predicts that for an agent with $n$ neighbors, punishment becomes a beneficial strategy when

(4)     $M > n.$

This is the biological expectation (compare to [13]).

In contrast, the economic expectation of behavior evolved through maximization of $p$ is that $i$ will never punish. As (2a) shows, $p_1 < p_0$ for any positive amount of punishment. Punishment is never a beneficial strategy when absolute payoff is what matters.

To compare these predictions, the simulation described above was run while systematically varying the punishment parameter $M$, and were conducted on the lattices listed in Table 1 to vary the number of neighbors $n$. 100 replications of the simulation were run for each value of $M$ starting at $M = 0.0$ and subsequently at increments of 0.5 until $M = 6.0$.

# 4 Results and Discussion

Results were mixed but generally favor the idea of the biological viewpoint that the evolution of agent behavior is driven by maximizing relative payoff.

## 4.1 Occurrence of Punishment

Experimental results are presented in Table 3. As $M$ increased, simulations run on all lattice networks eventually reached a value of $M$ at which the population underwent a rapid transition from defectors to cooperators. It is assumed in this study that punishment was the mechanism driving this flip to cooperative behavior and that the transition value of $M$ indicates the point at which punishment became a beneficial strategy. Only in simulations run on a complete network did cooperative behavior never evolve. This was true for complete networks even at $M = 5,000$.

**Table 3**. Simulation results versus economic and biological expectations ($N = 400$).

| Network Type | value of $M$ at which punishment becomes beneficial | | |
|---|---|---|---|
| | Economic expectation | Biological Expectation | Simulation Result |
| Ring | Never | 2 | 1.5 |
| Von Neumann | Never | 4 | 1.8 |
| Hexagonal | Never | 6 | 2.2 |
| Moore | Never | 8 | 2.8 |
| Complete | Never | $N - 1$ | Never |

This result suggests that the simple economic premise that agents maximize absolute payoff is not valid for populations embedded in discrete social structures. Only in a complete network, which is analogous to a well-mixed, homogeneous system, did simulation results match economic expectations. Instead results support the biological notion that, when sufficiently strong, punishment may become evolutionarily beneficial and, in turn, induce cooperative behavior in a population.

This result is especially significant given that human populations are not homogeneous and well-mixed, but are structured by complex interaction networks [14].

## *4.2 Punishment and number of neighbors*

On lattices where punishment behavior did evolve, the biological expectation was that punishment would proliferate when $M > n$ (4), where $n$ is the number of neighbors an agent has. Table 3 shows that for each lattice type, the value of $M$ at which punishment actually became prevalent was lower than predicted by this model[2].

This result indicates that subtle effects of the network structure are likely missing from the simplistic prediction of (4). If researchers and policy makers are to better predict behavior of rational agents in structured populations, they must not only re-evaluate their assumptions of rationality, they must also begin to understand and quantify these network effects.

One possible reason that punishment emerges at a lower value of $M$ than expected is that a cheater in agent $i$'s neighborhood is also a member of other neighborhoods and may be incurring punishment from agents other than $i$. In addition, neighbors of $i$ may be lowering their payoffs by engaging in their own altruistic punishment. Both of these events decrease $P_1$ in equation (3), leading to a lower transition value of $M$.

## 5 Summary

Many problems facing policy makers today require tools that facilitate the prediction of human behavior. Since many models of human behavior are premised on the concept of economic rationality, it is critical to evaluate the fundamental assumptions of rationality. Results of this study suggest that reassessing definitions of rationality from a biological perspective may lead to higher predictive accuracy. In particular, they suggest that the roles of relative and absolute payoff maximization should be reviewed when defining rationality.

Over 100 years ago economist Thorstein Veblen bluntly asked, "Why is economics not an evolutionary science?" [15]. The intent of this pilot study is to continue

---

[2] Interestingly, observed values approximate the square root of biologically expected values, though without further investigation and a greater number of data points, no speculation is made here regarding the validity or possible causality of this relationship.

Veblen's quest by opening a dialog between evolutionary biologists and behavioral economists. The ultimate goal of such a dialog is, by ensuring that any definition of rationality is firmly grounded in Darwinian theory, to improve both understanding and prediction of human behavior.

**Acknowledgements** I gratefully acknowledge Dr. Ann Kinzig for critical feedback and guidance related to this work. This material is based upon work supported in part by the National Science Foundation, through a Graduate Research Fellowship, and in part by an IGERT Fellowship in Urban Ecology from Arizona State University. Any opinions, findings, and conclusions or recommendations expressed in this material are those of the author and do not necessarily reflect the views of the National Science Foundation.

# References

1. Bowles S, Gintis H (2004) The evolution of strong reciprocity: cooperation in heterogeneous populations. Theor Popul Biol 65: 17-28
2. Boyd R, Gintis H, Bowles S et al (2003) The evolution of altruistic punishment. Proc Natl Acad Sci USA 100: 3531-3535
3. Fowler J H (2005) Altruistic punishment and the origin of cooperation. Proc Natl Acad Sci USA 102: 7047-7049
4. Andreoni J, Harbaugh W, Vesterlund L (2003) The carrot or the stick: Rewards, punishments, and cooperation. Am Econ Rev 93: 893-902
5. Gürerk Ö, Irlenbusch B, Rockenbach B (2006) The competitive advantage of sanctioning institutions. Science 312: 108-111
6. Fehr E, Fischbacher U (2004) Third-party punishment and social norms. Evol Hum Behav 25: 63-87
7. Fehr E, Gächter S (2000) Cooperation and punishment in public goods experiments. Am Econ Rev 90: 980-994
8. Fehr E, Gächter S (2002) Altruistic punishment in humans. Nature 415: 137-140
9. Ostrom E, Walker J, Gardner R (1992) Covenants with and without a Sword - Self-Governance Is Possible. Am Polit Sci Rev 86: 404-417
10. McAdams R H (1992) Relative Preferences. Yale Law J 102: 1-104
11. Shutters S T (2008) Strong reciprocity, social structure, and the evolution of fair allocations in a simulated ultimatum game. Computational Math Organ Theory, In Press, publ online 23-Oct-2008, DOI 10.1007/s10588-008-9053-z
12. Killingback T, Doebeli M (2002) The continuous prisoner's dilemma and the evolution of cooperation through reciprocal altruism with variable investment. American Naturalist 160: 421-438
13. Ohtsuki H, Hauert C, Lieberman E et al (2006) A simple rule for the evolution of cooperation on graphs and social networks. Nature 441: 502-505
14. Watts D J, Strogatz S H (1998) Collective dynamics of 'small-world' networks. Nature 393: 440-442
15. Veblen T (1898) Why is economics not an evolutionary science? Quart J Econ 12: 373-397

# A Network-Based Approach to Understanding and Predicting Diseases

Karsten Steinhaeuser[1] and Nitesh V. Chawla[2]

[1]ksteinha@cse.nd.edu, University of Notre Dame, IN, USA
[2]nchawla@cse.nd.edu, University of Notre Dame, IN, USA

**Abstract** Pursuit of preventive healthcare relies on fundamental knowledge of the complex relationships between diseases and individuals. We take a step towards understanding these connections by employing a network-based approach to explore a large medical database. Here we report on two distinct tasks. First, we characterize networks of diseases in terms of their physical properties and emergent behavior over time. Our analysis reveals important insights with implications for modeling and prediction. Second, we immediately apply this knowledge to construct patient networks and build a predictive model to assess disease risk for individuals based on medical history. We evaluate the ability of our model to identify conditions a person is likely to develop in the future and study the benefits of demographic data partitioning. We discuss strengths and limitations of our method as well as the data itself to provide direction for future work.

## 1 Introduction

Medical research is regarded as a vital area of science as it directly impacts the quality of human life. One prominent contribution of medicine is a steady increase in life expectancy [7, 8, 9]. However, advanced procedures are often expensive and required for longer periods due to extended lifetime, resulting in rising health care costs. Under the Medicare system, this burden falls primarily on society at large. Despite a general consensus that growth of health-related expenditures cannot continue indefinitely at the current rate[1], there are many competing plans for alleviating this problem. Most call on government, insurance companies, and/or employers to help individuals, but such plans simply re-distribute the burden and therefore only provide temporary relief. An alternate view suggests that the issue is not payment, but rather our general approach to health care. More specifically, the current system is reactive, meaning that we have become proficient at diagnosing diseases and developing treatments to cure or prolong life even with chronic conditions. In contrast, we should strive for *proactive personalized care* wherein susceptibility of an individual to conditions is assessed and preventive measures to counter high-risk diseases are taken. Since universal testing is cost-prohibitive, we must rely on generalized predictive models to assess disease risk [3, 10]. Such an approach relies on an understanding of the complex relationships between diseases in a population.

---

[1] http://www.whitehouse.gov/stateoftheunion/2006/healthcare/index.html#section2

H. Liu et al. (eds.), *Social Computing and Behavioral Modeling*,
DOI: 10.1007/978-1-4419-0056-2_26, © Springer Science + Business Media, LLC 2009

**Contributions**

The aforementioned issues are addressed through graph-based exploration of an extensive Medicare database. 1) We construct disease networks and study their structural properties to better understand relationships between them; we also analyze their behavior over time. 2) We describe a generalized predictive model that takes as input the medical history of an individual, extracts patient networks based on the concept of nearest neighbors, and provides a ranked list of other conditions the person is likely to develop in the future.

**Organization**

The remainder of this paper is organized as follows. The dataset is introduced in Section 2. In Section 3 we describe the disease networks and analyze their phyiscal characteristics. In Section 4 we present the predictive model and experimental results demonstrating the effects of data partitioning using demographic attributes. We conclude with a discussion of findings in Section 5, including an assessment of the data and methods used, as well as directions for future work.

# 2 Data

The database used in this study was compiled from raw claims data at the Harvard University Medical School and is comprised of Medicare beneficiaries who were at least 65 years of age at the time of their first visit. Data spans the years 1990 to 1993 and consists of 32 million records, each for a single inpatient visit, representing over 13 million individual patients. A record consists of the following fields: unique patient ID, date of admission and age of the patient at that time, the demographic attributes gender and ethnicity, the state in which the health care provider is located, and whether or not the patient's income lies below the poverty line.

In addition, a record contains up to ten different diagnosis codes as defined by the *International Classification of Diseases, Ninth Revision, Clinical Modification*[2] (ICD-9-CM). The first code is the principal diagnosis, followed by any secondary diagnoses made during the same visit. The number of visits per patient ranges from 1 to 155, and an average of 4.32 diagnosis codes are assigned per visit.

There were some obvious problems with the raw data such as dates outside the specified range, invalid disease specifications, and value-type mismatches, which we attribute to registration and transcription errors. We made our best effort to correct for these through extensive cleaning prior to starting our work or removing them if correction was not possible, but some noise inherently remains in this dataset as a result of misdiagnoses, incorrectly or incompletely entered codes, and the like.

---

[2] http://www.cdc.gov/nchs/about/otheract/icd9/abticd9.htm

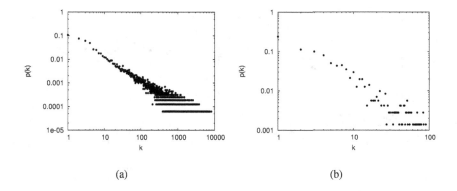

(a)                                                    (b)

**Fig. 1** Degree distribution for (a) the complete disease network constructed using five-digit codes (b) the network after collapsing nodes to three-digit codes and pruning at $w_{min} = 0.01$

## 3 Properties of the Disease Network

### Connecting Diseases

We use two related concepts to identify connections between diseases: *morbidity*, the number of cases in a given population; and *co-morbidity*, the co-occurence of two diseases in the same patient. We construct a network by linking all diseases $A, B$ that are co-morbid in any one patient and assigning edge weights $w(A, B)$ as follows.

$$weight(A, B) = \frac{Co\text{-}Morbidity(A, B)}{Morbidity(A) + Morbidity(B)}$$

Intuitively, the numerator gives higher weight to edges connecting diseases that occur together frequently, but the denominator scales back the weight for diseases that are highly prevalent in the general population. These tend to obscure unknown connections between less common diseases which are of primary interest here.

### Collapsing Nodes and Pruning Edges

ICD-9-CM defines a taxonomy of five-digit codes that enable a detailed designation of diseases and their causes. Using these to construct the network as described above results in thousands of nodes, millions of edges, and extremely dense connectivity, making interpretation quite difficult. However, the leading three digits of each code denote the general diagnosis, so in order to obtain a more meaningful network we *collapse* codes into three-digit nodes. Some information will be lost in the process, e.g. whether a broken arm was suffered during a fall or automobile accident, but such details are not of relevance here. Even with collapsed codes the network remains very dense, so we also *prune* all edges with weight below threshold $w_{min} = 0.01$ to eliminate links between diseases where one is very common (e.g. hypertension) and one is relatively rare as such connections contribute only limited information.

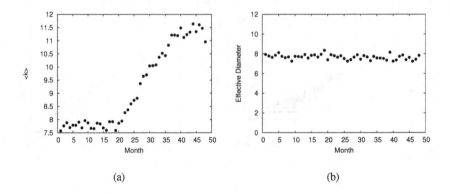

**Fig. 2** Plots of network properties (a) average node degree and (b) effective diameter by month.

**Network Characteristics**

After both collapsing and pruning, the network is reduced to a manageable size of 714 nodes and 3605 edges with average degree $\langle k \rangle = 10.1$, average clustering coefficient $C = 0.35$, and diameter of 14. Another fundamental network property is the degree distribution [1]. Figure 1(a) shows that the complete network is scale-free as the distribution follows a power law [2], indicated by the linear relationship in the log-log plot. By definition, this suggests that the collapsed and pruned network should also follow a power law, and indeed Figure 1(b) exhibits similar behavior.

The reduced network still has hubs and long-range edges one woulds expect in a scale-free network, but also a dense core dominated by diseases of the circulatory and digestive systems. In addition, several tight-knit communities form on the periphery roughly corresponding to diseases of the respiratory and genitourinary systems, as well as accidental injuries and poisoning. One unexpected observation is that neoplasms (cancer), known to spread throughout the body, do not form a community but rather occur with other diseases of the affected organs.

**Network Evolution**

Observing how properties change over time can also provide valuable information. For instance, [6] found that evolving networks tend to get more dense and as a result the diameter shrinks. We divide the data into one-month periods and apply the same methodology. Plots of average node degree and effective diameter for the network are shown in Figure 2. Given that the number of nodes remains approximately constant, the increasing trend in average degree indicates that the network does in fact become more dense over time. However, the effective diameter does not exhibit a discernible trend, which is surprising as one might expect some seasonal variability (e.g., infectious diseases like influenza) to be reflected in the network. This may imply that the relationships between diseases do *not* significantly vary with season or change over time, which in turn could improve predictive ability.

## 4 Disease Prediction in the Patient Network

We now address the task of predicting diseases that an individual is likely to develop in the future. To this end, we developed the *nearest neighbor network*, a network-based collaborative filtering method akin to those used for direct marketing (see [5] for a survey). However, here the network is based neither on explicit (direct communication) nor implicit (purchase of the same product) relations, but only on the (involuntary) sharing of some property – in this case a disease. The idea is to consider a limited medical history (i.e. small number of visits) and find other patients similar to the given person, who then "vote" on every disease the person has not yet had (based on their own medical histories). Votes are combined to produce a risk score for each disease. In this sense our approach is analogous to traditional nearest neighbor classification, but we include votes from more distant neighbors as well. The following section describes the method in detail.

### Nearest-Neighbor Networks

A nearest neighbor network is a hybrid between a nearest neighbor classifier and a collaborative filtering algorithm in that it selects the $k$ most similar entities to a target and then uses a weighted voting scheme to make predictions on the target.

*Patient Similarity* – Let a patient $P$ be defined by the set of diseases in his history, denoted by $diseases(P)$. A straighforward measure of the similarity $s$ between two patients $P, Q$ could then be defined as the number of diseases they have in common,

$$s(P, Q) = |diseases(P) \cap diseases(Q)|$$

The problem with this definition is that some patients have one hundred or more diseases in their medical history and are therefore similar to most other patients. To counter this effect we can use a quantity known as *Jaccard Coefficient* to compute similarity normalized by the total number of diseases two patients share,

$$s_{Jaccard}(P, Q) = \frac{|diseases(P) \cap diseases(Q)|}{|diseases(P) \cup diseases(Q)|}$$

Another problem is that some diseases are very common among this population, such as hypertension, heart disease, and urinary tract infections. Hence there exists an inherent bias in the data, resulting in some level of similarity between patients who otherwise share no medically meaningful conditions. In an attempt to correct for this bias we weight the contribution of each diseases $D$ to similarity by the inverse of morbidity (total number of patients with $D$) in the population,

$$s_{IFreq}(P, Q) = \sum_{D \in diseases(P) \cap diseases(Q)} \frac{1}{Morbidity(D)}$$

An experimental comparison of the latter two similarity measures showed little difference between them, so for the remainder of this work we use only $s_{Jaccard}$.

*From Neighbors to Networks* – In traditional nearest neighbor classification, we would simply consider the $k$ most similar other patients to make a prediction for some probe disease $D$. However, due to the sparsity of the data the amount of information provided by each neighbor may be limited (i.e. the probability of a neighbor having $D$ is low). Moreover, some patients only have a relatively small number of neighbors and hence increasing $k$ is ineffective. This situation is illustrated in Figure 3(a) where only a single neighbor has the probe disease, providing little confidence in the prediction.

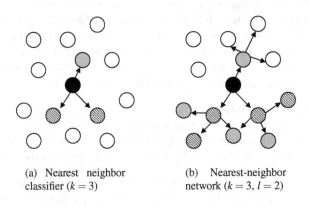

(a) Nearest neighbor classifier ($k = 3$)  (b) Nearest-neighbor network ($k = 3, l = 2$)

**Fig. 3** Comparison between traditional nearest neighbor classifier and nearest-neighbor network. The black node indicates the target patient, nearest neighbors are connected via directed edges. Neighbors who have the probe disease $D$ are shaded, those who do not are striped.

To overcome this limitation, we construct a network by finding the $k$ nearest neighbors of each patient and connecting them via directed edges. When probing for disease $D$, if an immediate neighbor does not have $D$ we recursively query this nearest neighbor network with depth-first search up to depth $l$. Figure 3(b) illustrates this process for the same example as before, but this time using the network with search depth $l = 2$. Four different patients with $D$ now contribute to the final score, increasing our confidence in the prediction.

**Experimental Setup**

To test the predictive model described above we select a subset of 10,000 patients, each of whom has at least five visits (to enable the validation against future visits). We construct the network with $k = 25$ using all visits for every patient. To make predictions for a patient we employ a hold-one-out method wherein we remove the corresponding node from the network, re-compute the patient's similarity to all others based only on the first three visits to find the 25 nearest neighbors, and re-insert the node into the network accordingly. We then iterate over all possible diseases probing the network for each, starting at the target patient, up to depth $l = 3$. A neighbor makes a contribution to the final score proportional to its similarity if he has had the disease, and none otherwise.

Our current model uses a linear decay for more distant neighbors, meaning that the contribution is divided by the depth of the node relative to the target patient. This process is repeated for all patients in the network.

## Results & Analysis

There is no straightforward quantitative methodology for evaluating the experimental results. We opted for a comparison-of-ranks method, which takes into account the prevalence of each disease in the population (also called *population baseline*). We begin by ranking all diseases based on this baseline. For each patient, we also rank the diseases according to the predicted risk score. We then search this ordered list for diseases the patient actually develops in future visits (i.e. those not used for computing similarity) and note whether they moved up, down, or remained in the same spot relative to the population baseline. A disease with a risk score of 0 did not receive any votes and is considered as missed.

The aggregate results over the full set of 10,000 patients is shown in the leftmost column of Table 1. We find that almost 42% of diseases moved in the desired direction (up), whereas nearly 30% moved down and another 26% were missed altogether. The diseases which moved down are problematic, but it might be possible to elevate their rank by using alternatve methods for weighting and combining votes. However, the large percentage of misses is of even greater concern as they cannot be found in the network at all.

**Table 1** Summary of experimental results predicting diseases using nearest neighbor networks, including a comparison of partitioning along different demographics (improvements in **bold**).

| Baseline Results | | Partitioned by Age | | Partitioned by Gender | | Partitioned by Race | |
|---|---|---|---|---|---|---|---|
| Up | 41.85% | Up | 39.29% | Up | **46.04%** | Up | 33.38% |
| Down | 28.57% | Down | 38.39% | Down | 40.43% | Down | 30.70% |
| Even | 3.23% | Even | 2.96% | Even | 2.97% | Even | 2.49% |
| Missed | 26.36% | Missed | **19.36%** | Missed | **10.57%** | Missed | 33.43% |

In an effort to improve performance of nearest neighbor networks for disease prediction, we incorporate an additional pre-processing step using data partitioning along demographic attributes. We divide the data as follows:

- Age - five-year bins starting at 65-69, 70-74, ..., 95-99, 100+.
- Gender - male and female.
- Race - grouped into six different codes; for privacy reasons, the actual ethnicity of each group could not be inferred from the data.

The nearest neighbor networks were then constructed within each partition and all predictions repeated. The results are also included in Table 1; favorable outcomes are shown in **bold**. Most notably, the fraction of diseases missed is significantly reduced when partitioning by age, and even more so by gender. However, most of the additional diseases found in the network moved down, requiring further analysis of the exact problem and possible solutions. The percentage of diseases that moved up only improved in the case of gender partitions, which makes sense as there are a number of diseases that are gender-specific.

# 5 Conclusion

Motivated by the ultimate goal of shifting medicine toward preventive health care, we performed a network-based exploratory analysis of an extensive medical dataset. Here we presented findings on two distinct tasks. First, we constructed a disease network and studied its properties. Specifically, we noted that while it is scale-free, it also contains discernible communities roughly corresponding to disease categories. Future work could compare this network based solely on observations to the genetic disease network built from phenotypic similarities between disease genomes [4]. Second, we introduced the concept of nearest-neighbor networks to assess disease risk for individual patients. We evaluated the use of data partitioning as a pre-processing step and found that demographics can improve coverage of predicted diseases, but only partitioning by gender provided actual improvements.

Based on our collective results, we believe that disease risk assessment is a promising research area. The ability to accurately predict future diseases could have a direct and profound impact on medicine by enabling personalized preventive health care. Additional work is required to minimize the number of diseases missed and maximize the ranking of correct diseases, but several variables have not been fully explored. For example, what is the optimal search depth? How much does each neighbor contribute to the final score, and what is the best method for combining the votes? There are also shortcomings with the data itself that must be addressed. For instance, to what extent can we reduce the effect of highly prevalent diseases? Does it even make sense to predict on all diseases? These questions will be answered through pursuit of higher quality data and improved predictive methods.

**Acknowledgements** Dr. Chawla is affiliated with the Center for Complex Network Research at the University of Notre Dame.

# References

1. R. Albert, A.-L. Barabási. *Statistical Mechanics of Complex Networks.* Rev. Modern Physics, 74, pp. 47–97, 2002.
2. A.-L. Barabási, E. Bonabeau. *Scale-Free Networks.* Scientific American, 288, pp. 60–69, 2003.
3. D. Benn, D. Dankel, and S. Kostewicz. *Can low accuracy disease risk predictor models improve health care using decision support systems?* AMIA Symposium, pp. 577–581, 1998.
4. K.-I. Goh, M. Cusik, D. Valle, B. Childs, M. Vidal, A.-L. Barabási. *The Human Disease Network.* PNAS, 104(21), pp. 8685–8690, 2007.
5. S. Hill, F. Provost, C. Volinsky. *Network-Based Marketing: Identifying Likely Adopters via Consumer Networks.* Stat. Sci., 21(2), pp. 256–276, 2006.
6. J. Leskovec, J. Kleinberg, C. Faloutsos. *Graph evolution: Densification and shrinking diameters.* ACM TKDD, 1(1), pp. 1–40, 2007.
7. C. Mathers, R. Sadana, J. Salomon, C. Murray, A. Lopez. *Healthy life expectancy in 191 countries, 1999.* The Lancet, 357(9269), pp. 1685–1691, 2001.
8. National Center for Health Statistics. *Health, United States, 2007, With Chartbook on Trends in the Health of Americans.* Hyatsville, MD, 2007.
9. J. Riley. *Rising Life Expectancy: A Global History* Cambridge University Press, 2001.
10. P. Wilson, R. D'Agostino, D. Levy, A. Belanger, H. Silbershatz,W. Kannel. *Prediction of Coronary Heart Disease Using Risk Factor Categories.* Circulation, 97, pp. 1837–1847, 1998.

# Status and Ethnicity in Vietnam: Evidence from Experimental Games

Tomomi Tanaka[†] and Colin F. Camerer[*]

[†] tomomi.tanaka@asu.edu, Arizona State University, Tempe, AZ
[*] camerer@hss.caltech.edu, California Institute of Technology, Pasadena, CA

**Abstract** A common simplification of economic theory assumes that people only care about maximizing their own material payoffs. However, charitable giving and recent findings in experimental economics demonstrate people will sometimes sacrifice their own payoffs to change the payoffs of others. Individuals tend to care about equity [4, 12], and try to maximize social welfare and help the least well off [1, 6, 8]. People are often willing to punish others who violate social norms even when punishment is costly to them [10, 11]. People frequently trust and reciprocate even when exchange is anonymous [2]. Thus, behavior seems to reflect a mixture of selfishness and "pro-sociality" toward non-kin to an extraordinary degree, compared to other species.

Pro-social preferences and cooperation are critical factors in facilitating economic exchange [7]. Pro-social preferences are also likely to be expressed in social policies, since political support for redistribution policies depends largely on whether the poor are perceived as deserving help and the rich are perceived as obligated to help. The framing of such policies was sharply contrasted in the 2008 U.S. presidential election. One side portrayed a proposed tax increase on the upper 5% of wage earners as a "patriotic" duty of the wealthy while the other side called the same proposal "class warfare".

However, pro-social preferences also appear to be affected by group membership. Social psychology research long ago showed that people usually are more pro-social toward in-group counterparts, even when "minimal" groups are created artificially and instantly (by dividing subjects into two rooms, for example) [15]. Anthropologists and theoretical biologists have argued that the capacity for in-group favoritism is a key ingredient in gene-culture evolution, which creates indirect group selection and explains the rapid progress of human civilization (and its failings, resulting in wars and genocides) [5].

A few recent experiments have shown interesting patterns of in-group favoritism in naturally-occurring groups in field settings [3, 9, 13]. We conducted experiments in Vietnamese village communities to investigate how social status, exemplified by ethnicity, affects preferences for distributions of income and economic cooperativeness. Experiments were conducted with three ethnic groups, Vietnamese (the majority), Chinese (a rich minority) and Khmer (a poor minority).

Our study extends these important results in three ways. First, isolating pure in-group favoritism in field settings requires controlling for covariates such as income, education, and occupation, which often vary strongly across groups. We control for these variables at the individual level using an extensive household survey. Second, social stereotypes about warmth and competence of groups have been conjectured to drive status judgments [14]. We administered established psychometric scales to each individual subject about each other group. Those judgments are used as covariates, to see if group favoritism effects are subsumed by social stereotyping. Third, we used a battery of five different games. Three games tap altruism (envy, dictator, and third-party punishment games) and two tap economic exchange (trust and coalition games). This design shows both similarities and differences across games and enables stronger inference since behavior in any one game is often noisy.

# References

1. Andreoni, J. and J. Miller (2002) Giving according to GARP: An experimental test of the consistency of preferences for altruism. Econometrica. 70(2): 737-753
2. Berg, J., J. Dickhaut, and K. McCabe (1995) Trust, reciprocity and social history. Games and Economic Behavior. 10: 122-142
3. Bernhard, H., U. Fischbacher, and E. Fehr (2006) Parochial altruism in humans. Nature. 442: 912-915
4. Bolton, G.E. and A. Ockenfels (2000) ERC: A Theory of equity, reciprocity, and competition. American Economic Review. 90(1): 166-193
5. Boyd, R. and P. J. Richerson (2005) The origin and evolution of cultures. Oxford University Press, Oxford
6. Charness, G. and M. Rabin (2002) Understanding social preferences with simple tests. Quarterly Journal of Economics. 117(3): 817-869
7. Durlauf, S.N. (2002) On the empirics of social capital. Economic Journal. 112(483): F459-F479
8. Engelmann, D. and M. Strobel (2004) Inequality aversion, efficiency, and Maximin preferences in simple distribution experiments. American Economic Review. 94(4): 857-869
9. Falk, A. and C. Zehnder (2007) Discrimination and in-group favorism in a citywide trust experiment. IZA and University of Bonn Working Paper
10. Fehr, E. and U. Fischbacher (2004) Third-party punishment and social norms. Evolution and Human Behavior 25: 63-87
11. Fehr, E. and S. Gachter (2000) Cooperation and punishment in public goods experiments. American Economic Review. 90(4): 980-994
12. Fehr, E. and K.M. Schmidt (1999) A theory of fairness, competition, and cooperation. Quarterly Journal of Economics. 114(3): 817-868
13. Fershtman, C. and U. Gneezy (2001) Discrimination in a segmented society: An experimental approach. Quarterly Journal of Economics. 116(1): 351-377
14. Fiske, et al. (2002) A model of (often mixed) stereotype content: Competence and warmth respectively follow from perceived status and competition. Journal of Personality & Social Psychology. 82(6): 878-902
15. Tajfel, H. (1970) Experiments in intergroup discrimination. Scientific American. 223: 96-102

# Behavior Grouping based on Trajectory Mining

Shusaku Tsumoto and Shoji Hirano

{tsumoto,hirano}@med.shimane-u.ac.jp, Shimane University, Izumo, Japan

**Abstract** Human movements in a limited space may have similar characteristics if their targets are the same as others. This paper focuses on such a nature of human movements as a trajectory in two or three dimensional spaces and proposes a method for grouping trajectories as two-dimensional time-series data. Experimental results show that this method successfully captures the structural similarity between trajectories.

## 1 Introduction

Human movements in a limited space may have similar characteristics if their targets are the same as others. For example, if a person who wants to buy some food with alcoholic drinks go to a supermarket, he/she may buy both drinks and food which will be matched to the drinks. If we take a close look at the nature of human movements in the supermarket, we may find more precise characteristics of the clients, such as chronological nature of their behavior, which may show that a royal customer tends to go to fish market before they go to buy beer.

One of the most simple measurements of human spatio-temporal behavior is to set up the two-dimensional coordinates in the space and to measure each movement as a trajectory. We can assume that each trajectory includes information about the preference of a customer. Since the collection of trajectories can be seen as a spacio-temporal sequence in the plane, we can extract the common nature of sequences after we apply clustering method to them. When we go back to the data with the clusters obtained, we may have a visualization of common spatio-temporal behavior from the data, which will be the first step to behavioral modeling.

This paper focuses on such a nature of human movements as a trajectory in two or three dimensional spaces and proposes a method for grouping trajectories as two-dimensional time-series data, consisting of the following two steps. Firstly, it compared two trajectories based on their structural similarity, determines the best correspondence of partial trajectories and calculates the dissimilarity between the sequences. Then clustering method are applied by using the dissimilarity matrix.

The remainder of this paper is organized as follows. In Section 2 we describe the methodoology, including preprocessing of the data. In Section 3 we demonstrate the usefulness of our approach through the grouping experiments of trajectories on hand language. Section 4 is a conclusion of this paper.

H. Liu et al. (eds.), *Social Computing and Behavioral Modeling*,
DOI: 10.1007/978-1-4419-0056-2_28, © Springer Science + Business Media, LLC 2009

# 2 Method

## 2.1 Multiscale Description of Trajectories by the Modified Bessel Function

Let us consider examination data for one person, consisting of $I$ different time-series examinations. Let us denote the time series of $i$-th examination by $ex_i(t)$, where $i \in I$. Then the trajectory of examination results, $c(t)$ is denoted by

$$c(t) = \{ex_1(t), ex_2(t), \ldots, ex_I(t)\}$$

Next, let us denote an observation scale by $\sigma$ and denote a Gaussian function with scale parameter $\sigma^2$ by $g(t, \sigma)$. Then the time-series of the $i$-th examination at scale $\sigma$, $EX_i(t, \sigma)$ is derived by convoluting $ex_i(t)$ with $g(t, \sigma)$ as follows.

$$EX_i(t, \sigma) = ex_i(t) \otimes g(t, \sigma) = \int_{-\infty}^{+\infty} \frac{ex_i(u)}{\sigma\sqrt{2\pi}} e^{\frac{-(t-u)^2}{2\sigma^2}} du$$

Applying the above convolution to all examinations, we obtain the trajectory of examination results at scale $\sigma$, $C(t, \sigma)$, as

$$C(t, \sigma) = \{EX_1(t, \sigma), EX_2(t, \sigma), \ldots, EX_I(t, \sigma)\}$$

By changing the scale factor $\sigma$, we can represent the trajectory of examination results at various observation scales. Figure 1 illustrates an example of multiscale representation of trajectories where $I = 2$. Increase of $\sigma$ induces the decrease of convolution weights for neighbors. Therefore, more flat trajectories with less inflection points will be observed at higher scales.

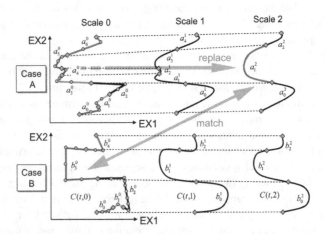

**Fig. 1** Multiscale representation and matching scheme.

Curvature of the trajectory at time point $t$ is defined by, for $I = 2$,

$$K(t, \sigma) = \frac{EX_1' EX_2'' + EX_1'' EX_2'}{(EX_1'^2 + EX_2'^2)^{3/2}}$$

where $EX_i'$ and $EX_i''$ denotes the first- and second-order derivatives of $EX_i(t, \sigma)$ respectively. The $m$-th order derivative of $EX_i(t, \sigma)$, $EX_i^{(m)}(t, \sigma)$, is defined by

$$EX_i^{(m)}(t, \sigma) = \frac{\partial^m EX_i(t, \sigma)}{\partial t^m} = ex_i(t) \otimes g^{(m)}(t, \sigma)$$

It should be noted that many of the real-world time-series data, including medical data, can be discrete in time domain. Thus, a sampled Gaussian kernel is generally used for calculation of $EX_i(t, \sigma)$, changing an integral to summation. However, Lindeberg [9] pointed out that, a sampled Gaussian may lose some of the properties that a continuous Gaussian has, for example, non-creation of local extrema with the increase of scale. Additionally, in a sampled Gaussian kernel, the center value can be relatively large and imbalanced when the scale is very small. Ref. [9] suggests the use of kernel based on the modified Bessel function, as it is derived by incorporating the discrete property. Since this influences the description ability about detailed structure of trajectories, we employed the Lindeberg's kernel and derive $EX_i(t, \sigma)$ as follows.

$$EX_i(t, \sigma) = \sum_{n=-\infty}^{\infty} e^{-\sigma} I_n(\sigma) ex_i(t - n)$$

where $I_n(\sigma)$ denotes the modified Bessel function of order $n$. The first- and second-order derivatives of $EX_i(t, \sigma)$ are obtained as follows.

$$EX_i'(t, \sigma) = \sum_{n=-\infty}^{\infty} -\frac{n}{\sigma} e^{-\sigma} I_n(\sigma) ex_i(t - n)$$

$$EX_i''(t, \sigma) = \sum_{n=-\infty}^{\infty} \frac{1}{\sigma} (\frac{n^2}{\sigma} - 1) e^{-\sigma} I_n(\sigma) ex_i(t - n)$$

## 2.2 Segment Hierarchy Trace and Matching

For each trajectory represented by multiscale description, we find the places of inflection points according to the sign of curvature. Then we divide each trajectory into a set of convex/concave segments, where both ends of a segment correspond to adjacent inflection points. Let $A$ be a trajectory at scale $k$ composed of $M^{(k)}$ segments. Then $A$ is represented by $\mathbf{A}^{(k)} = \{a_i^{(k)} \mid i = 1, 2, \cdots, M^{(k)}\}$, where $a_i^{(k)}$ denotes $i$-th segment at scale $k$. Similarly, another trajectory $B$ at scale $h$ is represented by $\mathbf{B}^{(h)} = \{b_j^{(h)} \mid j = 1, 2, \cdots, N^{(h)}\}$.

Next, we chase the cross-scale correspondence of inflection points from top scales to bottom scale. It defines the hierarchy of segments and enables us to guarantee the connectivity of segments represented at different scales. Details of the algorithm for checking segment hierarchy is available on ref. [1]. In order to apply the algorithm for closed curve to open trajectory, we modified it to allow replacement of odd number of segments at sequence ends, since cyclic property of a set of inflection points can be lost.

The main procedure of multiscale matching is to search the best set of segment pairs that satisfies both of the following conditions:

1. Complete Match: By concatenating all segments, the original trajectory must be completely formed without any gaps or overlaps.
2. Minimal Difference: The sum of segment dissimilarities over all segment pairs should be minimized.

The search is performed throughout all scales. For example, in Figure 1, three contiguous segments $a_3^{(0)} - a_5^{(0)}$ at the lowest scale of case $A$ can be integrated into one segment $a_1^{(2)}$ at upper scale 2, and the replaced segment well matches to one segment $b_3^{(0)}$ of case $B$ at the lowest scale. Thus the set of the three segments $a_3^{(0)} - a_5^{(0)}$ and one segment $b_3^{(0)}$ will be considered as a candidate for corresponding segments. On the other hand, segments such as $a_6^{(0)}$ and $b_4^{(0)}$ are similar even at the bottom scale without any replacement. Therefore they will be also a candidate for corresponding segments. In this way, if segments exhibit short-term similarity, they are matched at a lower scale, and if they present long-term similarity, they are matched at a higher scale.

## 2.3 Local Segment Difference

In order to evaluate the structural (dis-)similarity of segments, we first describe the structural feature of a segment by using shape parameters defined below.

1. Gradient at starting point: $g(a_m^{(k)})$
2. Rotation angle: $\theta(a_m^{(k)})$
3. Velocity: $v(a_m^{(k)})$

Figure 2 illustrates these parameters. Gradient represents the direction of the trajectory at the beginning of the segment. Rotation angle represents the amount of change of direction along the segment. Velocity represents the speed of change in the segment, which is calculated by dividing segment length by the number of points in the segment.

Next, we define the local dissimilarity of two segments, $a_m^{(k)}$ and $b_n^{(h)}$, as follows.

$$d(a_m^{(k)}, b_n^{(h)}) = \sqrt{\left(g(a_m^{(k)}) - g(b_n^{(h)})\right)^2 + \left(\theta(a_m^{(k)}) - \theta(b_n^{(h)})\right)^2}$$

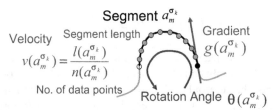

**Fig. 2** Segment Parameters.

$$+\left|v(a_m^{(k)}) - v(b_n^{(h)})\right| + \gamma\left\{cost(a_m^{(k)}) + cost(b_n^{(h)})\right\}$$

where $cost()$ denotes a cost function used for suppressing excessive replacement of segments, and $\gamma$ is the weight of costs. We define the cost function using local segment dissimilarity as follows. For a segment $a_m^{(k)}$ that replaces $p$ segments $a_r^{(0)} - a_{r+p-1}^{(0)}$ at the bottom scale,

$$cost(a_m^{(k)}) = \sum_{q=r}^{r+p-1} d(a_q^{(0)}, a_{q+1}^{(0)})$$

## 2.4 Sequence Dissimilarity

After determining the best set of segment pairs, we newly calculate value-based dissimilarity for each pair of matched segments. The local segment dissimilarity defined in the previous section reflects the structural difference of segments, but does not reflect the difference of original sequence values; therefore, we calculate the value-based dissimilarity that can be further used as a metric for proximity in clustering.

Suppose we obtained $L$ pairs of matched segments after multiscale matching of trajectories $A$ and $B$. The value-based dissimilarity between $A$ and $B$, $D_{val}(A,B)$, is defined as follows.

$$D_{val}(A,B) = \sum_{l=1}^{L} d_{val}(\alpha_l, \beta_l)$$

where $\alpha_l$ denotes a set of contiguous segments of $A$ at the lowest scale that constitutes the $l$-th matched segment pair $(l \in L)$, and $\beta_l$ denotes that of $B$. For example, suppose that segments $a_3^{(0)} \sim a_5^{(0)}$ of $A$ and segment $b_3^{(0)}$ of $B$ in Figure 1 constitute the $l$-th matched pair. Then, $\alpha_l = a_3^{(0)} \sim a_5^{(0)}$ and $\beta_l = b_3^{(0)}$, respectively. $d_{val}(\alpha_l, \beta_l)$ is the difference between $\alpha_l$ and $\beta_l$ in terms of data values at the peak and both ends of the segments. For the $i$-th examination $(i \in I)$, $d_{val_i}(\alpha_l, \beta_l)$ is defined as

$$d_{val_i}(\alpha_l, \beta_l) = peak_i(\alpha_l) - peak_i(\beta_l)$$

$$+\frac{1}{2}\{left_i(\alpha_l) - left_i(\beta_l)\} + \frac{1}{2}\{right_i(\alpha_l) - right_i(\beta_l)\}$$

where $peak_i(\alpha_l)$, $left_i(\alpha_l)$, and $right_i(\alpha_l)$ denote data values of the $i$-th examination at the peak, left end and right end of segment $\alpha_l$, respectively. If $\alpha_l$ or $\beta_l$ is composed of plural segments, the centroid of the peak points of those segments is used as the peak of $\alpha_l$. Finally, $d_{val_i}$ is integrated over all examinations as follows.

$$d_{val}(\alpha_l, \beta_l) = \frac{1}{I}\sqrt{\sum_i d_{val_i}(\alpha_l, \beta_l)}$$

## 2.5 Clustering

For clustering, we employ two methods: agglomerative hierarchical clustering (AHC) [13]. The sequence comparison part performs pairwise comparison for all possible pairs of time series, and then produces a dissimilarity matrix. The clustering part performs grouping of trajectories according to the given dissimilarity matrix.

## 3 Experimental Results

We applied the proposed method to the Australia sign language (ASL) dataset from UCI KDD archive [11]. The Australia sign language dataset contained trajectories on the movement of hands for 95 signs collected from five signers. The class (word) of each trajectory was known in advance; therefore, we used this dataset for quantitative performance evaluation and comparison with other methods.

The dataset contained time-series data on the hand positions collected from 5 signers during performance of sign language. For each signers, two to five sessions were conducted as shown in Table 1. In each session, one to six (mostly five) sign samples were recorded for each of the 95 words, and a total of 6757 sign samples were included in the dataset. The position of the hand was represented by x (left/right), y (up/down) and z (backward/forward), whose values were normalized in advance to [-1, 1]. The length of each sample was different and typically contained about 50-150 time points.

**Table 1** Summary of the ASL data

| Signer | Sessions | No of samples in each session (samples · words) | | | | |
|--------|----------|------------|------------|------------|------------|------------|
| Adam | 2 | 190 (2·95) | 570 (6·95) | | | |
| Andrew | 3 | 95 (1·95) | 475 (5·95) | 190 (2·95) | | |
| John | 5 | 95 (1·95) | 475 (5·95) | 475 (5·95) | 475 (5·95) | 190 (2·95) |
| Stephen | 4 | 98 (1·95+3) | 490 (5·95+15) | 489 (5·94 +19) | 490 (5·95 +15) | |
| Waleed | 4 | 490 (5·95+15) | 490 (5·95+15) | 490 (5·95+15) | 490 (5·95+15) | |

This dataset was used for experimental evaluation by Vlachos et al. [3] (used as 2-D trajectories) and Keogh et al. [12] (used as 1-D time series). We followed the experimental procedures described in [3]. Out of the 95 signs (words), the following 10 words were selected: *Norway, cold, crazy, eat, forget, happy, innocent, later, lose, spend*. The two-dimensional trajectories were constructed from $x$ and $y$ values in the dataset. We used the raw data without any filtering. The task of clustering was described as follows.

1. Select a pair of words such as {*Norway, cold*}. For each word, there exist 5 sign samples; therefore a total of 10 samples are selected.
2. Calculate the dissimilarities for each pair of the 10 samples by the proposed method.
3. Construct two groups by applying average-linkage hierarchical clustering [13] (split the data at the top branch of the dendrogram).
4. Evaluate whether the samples are grouped correctly.

The number of combinations for choosing a pair of words from ten of them was 45. The evaluation was made according to the number of correctly grouped pairs of words, e.g., 0/45 for the worst and 45/45 for the best.

Since we could not identify which signer/session's data were used in the previous works, we examined the data of all sessions that contain exactly five samples for the above 10 words. The list of sessions included: Andrew2, John2-4, Stephen1, Stephen4 and Waleed1-4 (denoted as Name+Session; for example, Andrew2 represents the data of the 2nd session for signer Andrew). We excluded Stephen2 because it contained only 4 samples for the word *happy*.

Parameters for multiscale matching were determined through a pilot experiment as followings: starting scale = 0.25, scale interval = 0,1, number of scales = 600, weight for segment replacement cost = 1.0.

Table 2 provides the results. The best result was obtained on the dataset Stephen2, with the correct clustering ratio of 0.844 (38/45). The worst result was obtained on the dataset Waleed3 with the ratio of 0.556 (25/45). According to [3], the results by the Euclidean distance, DTW, and LCSS were 0.333 (15/45), 0.444 (20/45), and 0.467 (21/45) respectively. Comparison of these results showed that our method could yield better results than others even for the worst case.

**Table 2** Clustering results of the ALS data

| Dataset | # of correctly grouped pairs | ratio | Dataset | # of correctly grouped pairs | ratio |
|---------|------------------------------|-------|---------|------------------------------|-------|
| Andrew2 | 26 / 45 | 0.578 | Stephen4 | 29 / 45 | 0.644 |
| John2 | 34 / 45 | 0.756 | Waleed1 | 33 / 45 | 0.733 |
| John3 | 29 / 45 | 0.644 | Waleed2 | 36 / 45 | 0.800 |
| John4 | 30 / 45 | 0.667 | Waleed3 | 25 / 45 | 0.556 |
| Stephen2 | 38 / 45 | 0.844 | Waleed4 | 26 / 45 | 0.578 |

# 4 Conclusions

In this paper we have presented a method for grouping trajectories. Our method employed a two-stage approach. Firstly, it compared two trajectories based on their structural similarity, and determines the best correspondence of partial trajectories. Then, it calculated the value-based dissimilarity for the all pairs of matched segments, and outputs their total sum as the dissimilarity of two trajectories. Experimental results on Australia sign language dataset demonstrated that our method could capture the structural similarity between trajectories even in the presence of noise and local differences, and could provide better proximity for discriminating different words.

# References

1. N. Ueda and S. Suzuki: A Matching Algorithm of Deformed Planar Curves Using Multiscale Convex/Concave Structures. IEICE Transactions on Information and Systems, J73-D-II(7): 992–1000 (1990).
2. X. Wang, A. Wirth, and L. Wang: Structure-Based Statistical Features and Multivariate Time Series Clustering. Proceedings of the 7th IEEE International Conference on Data Mining, 351–360 (2007).
3. M. Vlachos, G. Kollios and D. Gunopulos: Discovering similar multidimensional trajectories, Proceedings of the IEEE 18th International Conference on Data Engineering, 673–684 (2002).
4. J-G. Lee, J. Han, and K-Y Whang: Trajectory clustering: a partition-and-group framework. Proceedings of the 2007 ACM SIGMOD International Conference on Management of Data, 593–604 (2007).
5. A. P. Witkin: Scale-space filtering. Proc. the Eighth IJCAI, 1019–1022 (1983).
6. F. Mokhtarian and A. K. Mackworth: Scale-based Description and Recognition of planar Curves and Two Dimensional Shapes (1986). IEEE Transactions on Pattern Analysis and Machine Intelligence, PAMI-8(1): 24-43.
7. G. Dudek and J. K. Tostsos: Shape Representation and Recognition from Multiscale Curvature. Comp. Vis. Img Understanding, 68(2):170–189 (1997).
8. J. Babaud and A. P. Witkin and M. Baudin and O. Duda: Uniqueness of the Gaussian kernel for scale-space filtering (1986). IEEE Trans. PAMI, 8(1):26–33.
9. T. Lindeberg: Scale-Space for Discrete Signals. IEEE Transactions on Pattern Analysis and Machine Intellivence, 12(3):234–254 (1990).
10. Lowe, D.G: Organization of Smooth Image Curves at Multiple Scales. International Journal of Computer Vision, 3:119–130 (1980).
11. http://kdd.ics.uci.edu/databases/auslan/auslan.html
12. E. Keogh and M. Pazzani: Scaling up dynamic time warping for datamining applications. Proceedings of the sixth ACM SIGKDD International Conference on Knowledge Discovery and Data Mining, 285–289 (2000).
13. B. S. Everitt, S. Landau, and M. Leese: Cluster Analysis Fourth Edition. Arnold Publishers (2001).

# Learning to Recommend Tags for On-line Photos

Zheshen Wang and Baoxin Li

{zheshen.wang, baoxin.li}@asu.edu
Department of Computer Science and Engineering
Arizona State University, Tempe, AZ, USA

**Abstract**

Recommending text tags for on-line photos is useful for on-line photo services. We propose a novel approach to tag recommendation by utilizing both the underlying semantic correlation between visual contents and text tags and the tag popularity learnt from realistic on-line photos. We apply our approach to a database of real on-line photos and evaluate its performance by both objective and subjective evaluation. Experiments demonstrate the improved performance of the proposed approach compared with the state-of-the-art techniques in the literature.

## 1 Introduction

On-line services for archiving and sharing personal photos, such as Yahoo Flickr (www.flickr.com) and Google Picasa (picasa.google.com), have become more and more popular in recent years. Such services effectively provide a virtual social network or community for the users to share their memories, emotions, opinions, cultural experiences, and so on. Interaction among users in such a virtual community is largely enabled by information retrieval and exchange, e.g., photo retrieval and sharing, which are critically facilitated by tags or annotations of the photos. Tag recommendation systems (e.g., [1]) target at assisting users to come up with good tags that are both descriptive for a photo and useful for supporting information retrieval/exchange.

Photo tag recommendation is related to automated image annotation, which has received significant attention in recent years (e.g. [1-5]). For example, Duygulu et al. modeled annotation as machine translation [5], and Mori et al. used co-occurrence models [1]. In both cases, images are divided into sub-regions and a mapping between keywords and sub-regions are learnt. This and similar approaches are useful for images characterized by some key sub-regions but not so effective for generating tags with higher-level semantics that often link to the image as a whole. For example, given an image of the Great Wall, "China" may be one of commonly-used tags, which is unlikely to be directly predicted from purely visual features. Sigurbjornsson and Zwol proposed a tag co-occurrence algorithm for on-line photo tag recommendation [6],

H. Liu et al. (eds.), *Social Computing and Behavioral Modeling*,
DOI: 10.1007/978-1-4419-0056-2_29, © Springer Science + Business Media, LLC 2009

where the semantic relationship among tags is used for predicting new tags for a given image with some known tags. The limitation is that, the correlation analysis was purely based on text and thus at least one tag has to be present for the method to work.

Other related work includes collaborative filtering methods [7-9] and latent semantic analysis (LSA) [10-12]. Collaborative filtering is widely used in web recommendation systems for predicting (filtering) interests of a user through collecting information from many users, while LSA finds many applications in document comparison or retrieval in the concept space. While being useful, such methods in general do not readily extend to the processing of correlated visual and non-visual data as considered in this paper.

In this paper, we propose an automatic approach to tag recommendation for a given image without any annotations/tags. Our approach exploits the semantic correlation between image contents and text labels via Kernel Canonical Correlation Analysis (KCCA) [13]. In recommendation, the tags are ranked based on both the image-tag correlation and the input-independent tag popularity learnt from photos with user-created tags. We performed experiments using a realistic database collected on-line to demonstrate the superior performance of the proposed approach.

## 2 Semantic Image-Tag Correlation Analysis via KCCA

We propose to utilize Kernel Canonical Correlation Analysis (KCCA) to learn the underlying semantic correlation between the visual content and the textual tags of on-line photos and then use the correlation in predicting labels for images without tags. We briefly review the CCA/KCCA algorithm in the below.

CCA attempts to find basis vectors for two sets of variables such that the correlation between the projections of the variables onto these basis vectors is mutually maximized [14]. The canonical correlation between any two data sets is defined as

$$\rho = \max_{W_x, W_y} corr(F_x \cdot W_x, F_y \cdot W_y) \qquad (1)$$

where $F_x$ and $F_y$ are the two sets of variables, and $W_x$ and $W_y$ are the basis vectors onto which $F_x$ and $F_y$ are projected, respectively. This optimization problem can be formulated as a standard Eigen problem [13] which can be easily solved. There may be more than one canonical correlation, each representing orthogonally separate pattern of relationship between the two sets of variables. When extracting the canonical correlation the eigen values are calculated. The square root of the eigen values can be interpreted as the canonical coefficients. Corresponding to each canonical correlation the canonical weights for each of the variables in the data set are calculated. The canonical weights represent the unique positive or negative contribution of each variable to the total correlation.

CCA has been used previously by researchers to find the semantic relationship be-tween two multimodal inputs. In [15], CCA is used to find the language independent semantic representation of a text by using the English text and its French translation as set of variables. In this work, we model the semantic relationship between visual features of an image and the textual annotations/tags through CCA, and use available image features for predicting semantically-related texts in tag recommendation.

CCA is only able to capture linear correlations, while the actual correlation model in our application may be highly nonlinear. Therefore, we resort to the Kernel CCA, which has been used, e.g., in [13] for finding the correlation between image and text features in image retrieval. KCCA projects the data into a higher-dimensional feature space before performing CCA in the new feature space [13]:

$$\phi : x = (x_1, ..., x_m) \mapsto \phi(x) = (\phi_1(x), ..., \phi_N(x)), \quad (m < N) \tag{2}$$

where $x = (x_1, ..., x_m)$ is a set a variables and $\phi$ is a mapping from $m$ dimensions to $N$ dimensions. In this paper, we use the implementation made available by the authors of [13], with a Gaussian kernel.

# 3 Tag Recommendation: the Proposed Approach

We formulate tag recommendation as a tag ranking problem. We propose a novel ap-proach for ranking all possible tags by weighting image-tag correlation and input-independent tag popularity. An image-tag correlation score is obtained from KCCA and a tag popularity score is defined as the normalized tag frequency in the training set. The weights for the two scores can be used to control the relative contributions from the two parts. Details of the approach are presented in the below.

## 3.1 Tag Ranking

We first collect a training set of photos with corresponding tags. The vocabulary of possible tags is defined as a collection of all tags appearing in the training set: $\{t_j\}, j = 1, ..., m$. We define a ranking score for each tag in the vocabulary as below:

$$S_{t_j} = (1-a) \cdot S_{t_j}^{corr} + a \cdot S_{t_j}^{pop} \tag{3}$$

where $t_j$ is a possible tag from the vocabulary; $S_{t_j}^{corr}$ and $S_{t_j}^{pop}$ denote the semantic image-tag correlation score and the tag popularity score respectively, and $a \in [0,1]$ is a constant for weighting these two scores.

This new approach has some advantages, compared with our previous work [16] along the same direction. For example, all existing tags from the training set can be recommended, while in [16], only tags which are selected as keywords and are among the tag lists of the top correlated images may be chosen. In addition, the weight $a$ provides a flexible control of the contributions from the two terms, allowing for example the recommendation to rely more on visual contents or on the tag popularity.

## 3.2 Semantic Image-Tag Correlation Score

We compute the image-tag correlation score $S^{corr}$ based on KCCA as

$$S_{t_j}^{corr} = \max_{i=1...n} \{corr_{I_{t_j}^i}\} \tag{4}$$

where $t_j$ is a tag from the vocabulary of all possible tags; $I_{t_j}^i$ is the i[th] training instance (an image with corresponding tags) that $t_j$ belongs to, and $corr_{I_{t_j}}$ is the normalized correlation coefficient between this instance and the input instance. Conceptually, we use the maximum correlation coefficient between the input image and one of the training instances as the semantic image-tag correlation score (correlation coefficients are normalized in Eq. (4)). In our experiments, definitions of text and image features are adopted from our previous work [16]. For text features, we use the bag-of-word model in our experiments. Since our experiments are category related, we select keywords by adopting the *TFICF* weighting scheme [17]. For image features, we select spatial-pyramid-based HSV histograms plus Gabor gradients, which are able to describe both global and local color and spatial information of images [16]. We end up having a 756-d vector as the visual feature and a 252-d vector as the tag feature. The proposed algorithm for computing the image-tag correlation score is presented in the following.

### 3.2.1 Calculation of Instance-level Correlation Score

Assume that we have a training set which contains $n$ instances of images with corresponding tags. The training procedure is described as follows:

1. Extract text and image features from all instances in the training set and form feature matrices $F_x = [f_x^1, ..., f_x^n]^T$ and $F_y = [f_y^1, ..., f_y^n]^T$, where each row represents the feature vector of one instance.
2. Project $F_x$ and $F_y$ to a higher dimensional space by kernel mapping. The projected feature matrices are denoted as $F_x'$ and $F_y'$.
3. Perform KCCA between $F_x'$ and $F_y'$ and find the basis vectors:

$$[W_x, W_y] = KCCA(F_x', F_y') \tag{5}$$

in which $W_x$ and $W_y$ are the found KCCA basis matrices.
4. Project $F_x'$ and $F_y'$ onto the obtained basis:

$$F_x'' = F_x' \times W_x^k \tag{6}$$

$$F_y^{''} = F_y^{'} \times W_y^{k} \tag{7}$$

where $W_x^{k}$ and $W_y^{k}$ are obtained by selecting top $k$ basis vector from $W_x$ and $W_y$ respectively.

In the test stage, given a new test image, we first extract its image feature $f_{y_0}$ and obtain $f_{y_0}^{'}$ by projecting $f_{y_0}$ to a higher dimensional space using the same kernel mapping as used in Step 2 of the training procedure. We further project $f_{y_0}^{'}$ onto $W_y^{k}$ as we did for all training images and get $f_{y_0}^{''}$ as a result. Then, the instance level correlation score $corr_{I^i}$ of this image to the training instance $i$ can be computed as a normalized Pearson correlation between $f_{y_0}^{''}$ and $f_{y_i}^{''}$:

$$corr_{I^i} = \frac{correlation(f_{y_0}^{''}, f_{y_i}^{''})}{\max_{i=1...n}(correlation(f_{y_0}^{''}, f_{y_i}^{''}))} \tag{8}$$

where $f_{y_i}^{''}$ is the $i^{th}$ row of $F_y^{''}$.

### 3.3 Tag Popularity Score

The tag popularity score is an input-independent score for tags, which describes how likely a word is used as a tag based on the training set. This is defined as:

$$S_{t_j}^{pop} = c_{t_j} / \max_{k=1...m}\{c_{t_k}\} \tag{9}$$

where $t_j$ is a tag from the vocabulary of all possible tags and $c_{t_j}$ indicates the counts of appearances of $t_j$ in the training set. $S_{t_j}^{pop}$ is always within $[0,1]$.

## 4 Experiments and Results

According to a Yahoo study [6], the most frequent classes of tags for Yahoo Flickr photos are *locations, artifacts/objects, people/groups* and *actions/events*. In our work, we selected two popular topics for each class except *people/groups* (which will be included in future study due to its complexity). Specifically, we picked "office" and "stadium" as *location*, "pyramid" and "Greatwall" for *artifacts/objects*, and "skiing" and "sunset" *for actions/events*. For each topic, we collected 300 images from Flickr using the FlickrAPI tool. In order to mitigate user bias (images from the same user are visually very similar in many cases), we collected no more than 15 images from the same Flickr ID. For each topic, 200 images were used for training and 100 images for testing. The training and testing sets were defined by random selections (Since the 1800 images we collected are actually from 993 different Flickr IDs, there are only 2 images from the same user on average. Thus, random selection based on images is not much different from that based on their sources). Obviously, real on-line data is more challenging than research databases (e.g. the databases used in [4]) due to the varying sources of images and the uncontrollable usage of vocabulary in the user-provided tags. In order to show the improvements of the proposed approach, we use the same dataset which was used in our previous work [16] and keep top 15 tags as final recommendations.

### 4.1 Evaluation metrics

Both objective evaluation and subjective evaluation have been performed for validating and assessing the proposed method. For objective evaluation, we compared the recommended tags generated by our approach to the tags provided by the original owners of the photos. If one of the user tags is among the recommended tag list, we call it a *hit*. And we use *≥k-HitRate* for showing the performance, which gives the percentage of images out of all test images that achieve $\geq$ k *hit*.

For subjective evaluation, human evaluators were asked to visually check the images and mark on those tags which they deem as semantically relevant. In addition to *≥k-HitRate*, we also adopted the following statistical metrics from [6] for evaluating the performance: *Mean Reciprocal Rank (MRR)*, which measures where in the ranking the first relevant tag occurs; *Success at rank k (S@k)*, defined as the probability of finding a relevant tag among the top k recommended tags; *Precision at rank k (P@k)*, defined as the proportion of retrieved tags that is relevant, averaged over all photos.

### 4.2 Results and analysis

We ran ten trials with random selections for the training and test sets in order to avoid data selection bias. All experiments were based on these ten trials from the dataset.

Figure 1 and Table 1 show the results of objective evaluation of the proposed approach when *a* is set as 0.2 and 0.5 respectively. For k=1 and 2 cases, both achieved a hit rate close to 100% and above 70% respectively, which are superior to what we achieved in [16] (there is no objective evaluation results reported in [6]). For k=1 case, our result is even better than that in [4], with such a much more challenging dataset with high sparsity in tag occurrence. For each selected topic, only a few words appear more than 5 times in the user-provided tags for all the training images. This explains why the rate becomes lower when k increases to 2 and 3.

Objective evaluation alone cannot sufficiently evaluate the real performance since many recommended tags are actually good choices for tagging the images although the original users did not use them. If users are offered those recommended tags, they may likely select and use these tags. This is exactly what our tag recommendation system targets. Therefore, a subjective evaluation is necessary.

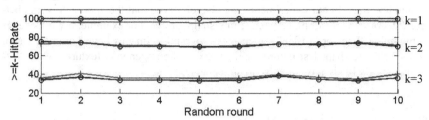

**Figure 1**: Comparisons of $\geq k$-*HitRate* between different field ranking methods based on objective evaluations: red cross--a=0.2; blue circle--a=0.5.

**Table 1** Tag hit rate of objective evaluation on test sets

| $\geq$k-HitRate (%)<br>Average over 10 fixed random rounds used in [16] | k=1 | k=2 | k=3 |
|---|---|---|---|
| a=0.2 | 97.0 | 71.8 | 37.1 |
| a=0.5 | 99.9 | 71.8 | 34.7 |

In subjective evaluations in this work, three participants were asked to tick all relevant tags in the recommended list for a given image. In order to avoid evaluator bias, each evaluator evaluates only 4 topics (400 test images) from the same random set and only two topics from the same user can be used in one user set. Thus we can have two user sets for this random test set. Except the $\geq k$-*HitRate* metric, we also employ MRR, S@1-5, P@5, P@10 and P@15 metrics as well. Average results for one of the ten random sets under different metrics are listed in Tables 2 and 3.

**Table 2.** Tag hit rate of subjective evaluation on one of the test sets.

| $\geq$k HitRate (%), a=0.5 | k=1 | k=2 | k=3 | k=4 | k=5 |
|---|---|---|---|---|---|
| User Set 1 | 99.2 | 86.2 | 64.3 | 44.0 | 26.2 |
| User Set 2 | 99.8 | 88.7 | 72.5 | 48.5 | 25.0 |

**Table 3.** Subjective evaluation on one of the test sets.

| a=0.5 | MRR | S@1 % | S@2 % | S@3 % | S@4 % | S@5 % | P@5 % | P@10 % | P@15 % |
|---|---|---|---|---|---|---|---|---|---|
| User Set 1 | 1.87 | 71.5 | 81.7 | 88.3 | 91.8 | 93.2 | 64.0 | 35.7 | 23.8 |
| User Set 2 | 1.66 | 78.3 | 84.8 | 90.5 | 93.7 | 95.3 | 66.9 | 35.2 | 23.5 |

The subjective evaluation results are statistically better than those of the objective evaluation, which supports our previous argument that, although many generated tags are not listed by the original user, they are good recommendations for the given image. Compared with the state-of-the-art performance in [6], our result is better than the best cases reported. Further, in [6], tag recommendation is purely based text and thus at least one tag from the user must be available; while in our experiment, tags can be recommended based on only images.

Both objective and subjective evaluation results demonstrate that the proposed approach is capable of capturing the underlying semantic correlation between image contents and text tags. In our experiments, we also observed that the approach works better for categories with strong coherence in visual features. For example, it works better for "pyramid" than for "office", since most "pyramid" images have common visual features (triangle-shaped structures and relatively homogeneous texture) while "office" images are likely more diverse in appearance.

## 5 Conclusion and Future Work

We propose a novel approach for tag recommendation for on-line photos, in which tag recommendation is formulated as a tag ranking problem. All tags from a training set are ranked by a weighted combination of semantic image-tag correlation and tag popularity learnt from the training set. Experimental results based on realistic on-line photos demonstrated the feasibility and effectiveness of the proposed method.

There are many other aspects that can be taken into consideration for further improving the work. For example, other available information of photos, such as title, description, comments, meta-data, etc., can be added as separated features for making tag recommendations, and the image-tag correlation score can be computed by combining the pair-wise top correlated instances obtained using these features. In addition, performing semantic grouping on tags before creating the document-term matrix, combining tag co-occurrence strategies proposed in [6] and analyzing user tagging history for providing customized recommendations are also promising directions. We will pursue those aspects in our future work.

## References

[1]  Y. Mori, H. Takahashi, and R. Oka, "Image-to-word transformation based on dividing and vector quantizing images with words," presented at International Workshop on Multimedia Intelligent Storage and Retrieval Management, 1999.
[2]  T. Kolenda, L. K. Hansen, J. Larsen, and O. Winther, "Independent component analysis for understanding multimedia content," presented at IEEE Workshop on Neural Networks for Signal Processing XII, 2002.
[3]  K. Barnard, P. Duygulu, N. d. Freitas, D. Forsyth, D. Blei, and M. I. Jordan, "Matching Words and Pictures," *Journal of Machine Learning Research*, vol. 3, pp. 1107-1135, 2003.
[4]  J. Li and J. Z. Wang, "Real-Time Computerized Annotation of Pictures," presented at ACM MM, Santa Barbara, USA, 2006.

[5] P. Duygulu, K. Barnard, N. d. Fretias, and D. Forsyth, "Object recognition as machine translation: Learning a lexicon for a fixed image vocabulary," presented at European Conference on Computer Vision (ECCV), 2002.

[6] B. Sigurbjornsson and R. v. Zwol, "Flickr Tag Recommendation based on Collective Knowledge," presented at ACM WWW2008, Beijing, China, 2008.

[7] P. Resnick, N. Iacovou, M. Suchak, P. Bergstrom, and J. Riedl, " GroupLens: An open architecture for collaborative filtering of netnews," presented at ACM Conference on Computer Supported Cooperative Work, Chapel Hill, NC, 1994.

[8] N. D.M., "Implicit Rating and Filtering," presented at Fifth DELOS Workshop on Filtering and Collaborative Filtering, Budapest, Hungary., 1997.

[9] C. W.-k. Leung, S. C.-f. Chan, and F.-l. Chung, "A collaborative filtering framework based on fuzzy association rules and multiple-level similarity," *Knowledge and Information Systems*, vol. 10, pp. 357-381, 2006.

[10] S. Deerwester, S. Dumais, G. W. Furnas, T. K. Landauer, and R. Harshman, "Indexing by latent semantic analysis," *Journal of the Society for Information Science.*, vol. 41, pp. 391-407, 1990.

[11] T. K. Landauer, D. S. McNamara, S. Dennis, and W. Kintsch, *Handbook of Latent Semantic Analysis*: Psychology Press, 2007.

[12] T. K. Landauer, P. W. Foltz, and D. Laham, "Introduction to Latent Semantic Analysis," *Discourse Processes*, vol. 25, pp. 259-284, 1998.

[13] D. R. Hardoon, S. R. Szedmak, and J. R. Shawe-taylor, "Canonical correlation analysis: An overview with application to learning methods," *Neural Computation*, vol. 16, pp. 2639-2664, 2004.

[14] H. Hotelling, "Relations between two sets of variates," *Biometrika*, vol. 28, pp. 312-377, 1936.

[15] A. Vinokourov, J. Shawe-Taylor, and N. Cristianini, "Inferring a semantic representation of text via cross-language correlation analysis," presented at NIPS, 2002.

[16] S. Subramanya, Z. Wang, B. Li, and H. Liu, "Completing Missing Views for Multiple Sources of Web Media," *International Journal of Data Mining, Modelling and Managment (IJDMMM)*, vol. 1.

[17] N. Agarwal, H. Liu, and J. Zhang, "Blocking objectionable web content by leveraging multiple information sources," *SIGKDD Explor. Newsl.*, vol. 8, pp. 17-26, 2006.

# A Social Network Model of Alcohol Behaviors

Edward J. Wegman[1] and Yasmin H. Said[2]

[1] ewegman@gmail.com, George Mason University, Fairfax, VA
[2] yasid99@hotmail.com, Gorge Mason University, Fairfax, VA

## 1 Introduction

Alcohol use is a major cause of morbidity and mortality in all areas of the world (Ezzati et al., 2002). In the developed world alcohol consumption is one of the primary risk factors for the burden of disease; it ranks as the second leading risk factor (after tobacco) for all disease and premature mortality in the United States (Said and Wegman, 2007). Alcohol is such a potent risk factor because it is widely used, inexpensive, and causes both acute and chronic consequences. A major difficulty in developing and assessing alcohol-related public health interventions is that drinking behaviors and their consequences form, in effect, a complex ecological system of individual behaviors embedded in socio-cultural settings and interactions that are distributed over space and time. Interventions focused on any specific aspect of this system (e.g., drinking and driving) may result in unintended consequences (e.g., a potential increase in domestic violence as the result of high volume drinkers spending more time at home due to the anti-drinking-driving programs). Another example of unintended consequences is alcohol-related promiscuous behavior resulting in infections with sexually transmitted diseases (STD) and Human Immunodeficiency Virus (HIV).

The research that we report here is designed to model simultaneously acute alcohol-related outcomes among all individuals by specific geographic areas. Thus, the potential impact of interventions that change any aspect of the entire alcohol ecological system can be examined across population subgroups and locations. This information can be used to guide the development of interventions that are most likely to have the impact intended, and to assist in the selection of interventions that are most worthy of deserving the large amount of resources necessary for conducting a public health field trial.

We develop a model of the alcohol ecological system based on population subgroups, age, gender, ethnicity, socioeconomic status, composed of individuals (i.e., the agents in the model), embedded in space (geographic location identified by postal code areas), time (i.e., time-of-day, day-of-week), and socio-cultural context (e.g., alcohol availability, neighborhood). The model focuses on acute alcohol-related outcomes (e.g., domestic violence, homicide, crashes, STD, HIV, etc.). The model is agent-based, meaning that individual outcomes are simulated; this software design allows us to incorporate specific agent interactions. This approach focuses on

H. Liu et al. (eds.), *Social Computing and Behavioral Modeling,*
DOI: 10.1007/978-1-4419-0056-2_30, © Springer Science + Business Media, LLC 2009

ecological systems, i.e., behaviors and interactions of individuals and their setting, not on aggregate groups as the unit of analysis.

The specific intent for this research is: 1) To create a spatiotemporal social network stochastic directed-graph model of a population accounting for agents' alcohol use and the associated acute outcomes; and 2) To incorporate a policy tool in the model in order to examine intervention strategies to understand where policies may be most effective in suppressing simultaneously all of the acute outcomes. This initial policy tool allows for examinations of the impact of changes in alcohol availability (by outlet type and by geographic location) and population demographics on the number, type, and location of acute alcohol-related outcomes.

## 2 Building the Model

Our model of the alcohol system is an agent-dependent, time-dependent stochastic digraph. The concept is that the vertices of the digraph will represent the state of the agent (e.g. including such factors as physical location, present activity, and level of Blood Alcohol Content (BAC)) and the edges represent a decision/action that takes the agent into a new state. The agent represents any individual in the population including the alcohol users as well as the non-users. The edge going from one state to another will have a conditional probability attached to it, hence, the notion of a stochastic digraph. The conditional probability attached to a given edge will depend on the specific sub-population from which the agent is drawn, hence is agent-dependent, and the conditional probability will, in principle, also depend on the time-of-day, hence it is time-dependent.

Implicit in this description is that individuals represented in the network relate to each other and to social institutions. Thus, overlaying the directed graph model is a social network. Figure 1 is a representation of the social network for the alcohol user. We develop an agent-based social network digraph model of the alcohol system. Implicit in the social network is that each node (each agent) carries with him or her covariates that characterize such factors as residence, ethnicity, job class, gender, age, socioeconomic status, disposition to alcohol, and the like.

The concept is that relatively homogeneous clusters of people (agents) will be identified along with their daily activities. Agents can be alcohol users, nonusers, and the whole suite of participants in the social network mentioned above. Agents can interact with each other in asymmetric ways. Their activities will be characterized by different states in the directed graph, and decision resulting in actions by an agent will move the agent from state to state in the directed graph. The leaf nodes in the graph will represent a variety of outcomes, most of which are benign, but a number of which will be acute alcohol-related outcomes. Specifically, we study simultaneously the following acute outcomes: assault, suicide, domestic violence, child abuse, sexual assault, STD and HIV, murder, DWI (drinking and driving episodes, including fatalities).

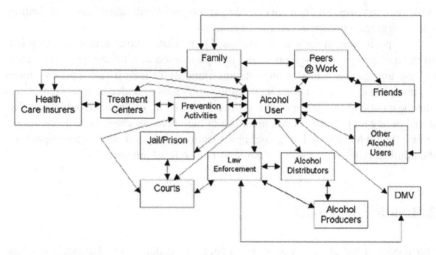

**Fig. 1** Social network for the alcohol user. This same network may apply to nonusers with some edges removed.

Figure 2 represents the essentially complete model. In each of the "time planes" we have the social network along with the attendant covariates. The conditional probabilities are, of course, time and agent dependent. We represent time as being quantized although it need not essentially be so. Indeed, while gender, ethnicity, job class are inherently categorical, such covariates as age, socioeconomic status as represented by income level and other covariates may be continuously variable. In the next section we will address compromises made related to data available to calibrate the data.

Now, as agents are introduced into the directed graph model, their ultimate outcomes whether benign or acute will be accumulated so that a (multinomial) probability distribution over all outcomes can be estimated. The ultimate goal is to introduce interventions that will alter the structure of the directed graph or its associated probabilities and assess how those interventions affect the probability distribution over the outcomes.

## 3 Choosing the Agents

Calibrating the model described in Section 2 would ideally involve calculating the conditional probability of each acute event conditioned on residence location, ethnicity, job class, gender, age, economic status, time-of-day, alcohol availability, whether the individual had an alcohol abuse problem or not, and conceivably many more factors that would affect the likelihood of that acute outcome. Unfortunately, there is no direct data to support estimation of the conditional probabilities. Part

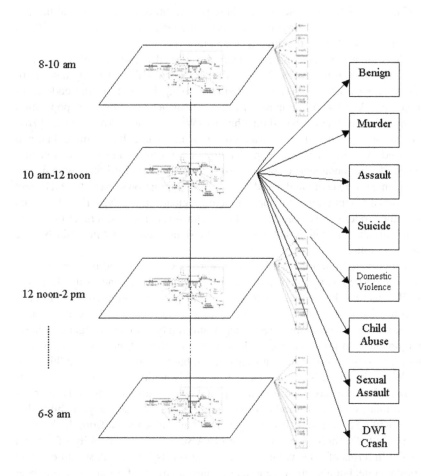

**Fig. 2** Time-of-day directed graph with acute outcomes. For illustrative purposes, we have blocked the day into twelve two-hour blocks. Note that any time of day, all outcomes are possible, but because drinking patterns are time-dependent, the likelihood changes throughout the day.

of the reason, therefore, for turning to a directed graph model is to decompose the desired probabilities into components that can be justified by available data.

The first goal is to devise a procedure that chooses agents in a manner that is representative of and consistent with the population of the geospatial area of interest. Note that if we considered only the factors mentioned above and quantized them at reasonable levels, say residence location: 50 postal codes in an area, ethnicity: 5 ethnic classes, job class: 3 broad job classes, gender: 2, age: 10 age blocks, socioeconomic status: 10, time-of-day: 12 two-hour blocks, alcohol availability: perhaps 3, and alcohol abuse problem or not: 2, we could be seeking to draw agents from 360,000 categories. And, of course, for each of these 360,000 categories, we would have to know the conditional probability of each of the eight acute events we are attempting to model. If we knew all of these probabilities, modeling could be captured by a relatively simple Bayesian network. Of course, we don't have all of these probabilities, so we turn to the directed graph model.

For illustrative purposes and admittedly because of data availability, we have chosen to use data from Fairfax County, Virginia, USA. Fairfax County has a population of approximately 1,000,000 individuals living in the 48 postal code areas in the county. As an approximation, we use a manageable number of population subgroups (classes) that are based on ethnicity (White, African American, and Hispanic) and job type (based on Census job codes, i.e. blue collar, white collar, and unemployed). For these classes, the probability of being an alcohol misuser (alcohol abuse and alcohol dependence) or nonmisuser (abstainer and drinkers who do not have a diagnosis) are estimated based on data from population surveys for each class (e.g., National Longitudinal Alcohol Epidemiological Survey, NLAES). Note that although the model is based on classes, the stochastic approach simulates outcomes for each individual, meaning that persons in the same class do not necessarily have the same outcome.

We know the population distribution in Fairfax County based on Census data. The simulation begins by choosing an agent at random from one of the 48 postal codes. Because we know the distribution of job class and ethnicity within each postal code also from U.S. Census data and data from the U.S. Bureau of Labor Statistics, we can attach to the selected agent information about the agent's ethnicity and job class. In principle, we could also attach information about age, gender, and socioeconomic status based on Census data, but we will not do so in this illustration. While the NLAES data does not provide information on whether someone is an alcohol misuser or not conditioned on postal code, it does provide this information conditioned on ethnicity and job class. This we can estimate the probability that our agent will be a misuser of alcohol. We also have information from the Virginia Department of Alcoholic Beverage Control (ABC) on the availability of alcohol principally in terms of sales volumes of distilled spirits for each ABC store within each postal code, but also more generically sales tax figures for outlets selling beer and wine. Thus, we can estimate the alcohol availability by postal code. It is known from studies such as Gorman et al. (2001) and Gruenewald et al. (2002) that alcohol availability directly influences drinking behaviors enhancing the probabilities of acute outcomes.

The prevalence of acute outcomes is dependent on the nature of the acute outcome, murder and suicide are fairly rare, DWI and assaults are fairly common. However, we have obtained annual aggregate figures in each category for Fairfax County for a period from 1999-2005. Finally, being an alcohol misuser can enhance the probability of engaging in violence by a factor of five or more (Collins and Schlenger, 1988 and Leonard et al., 2003). Thus, except for the time-of-day component, we can estimate the probability that our agent will engage in some acute outcome, although we have no information on which specific acute outcome. However, because we do know the distribution of acute outcomes annually, we can estimate by proportion the probability that our agent for each acute or benign outcome. This process will associate with each agent a likelihood of specific type of acute outcome and simultaneously give us aggregated information for each postal code, job class, ethnic type, and so on. Thus, for purposes of our approximation, we compute the following probability:

$$P(Acute\,Event_i|Agent_j) =$$

$$P(Z)P(E,J|Z)P(M|E,J)P(A|Z)P(Some\,Acute\,Event|M)P(Acute\,Event_i)+$$

$$P(Z)P(E,J|Z)P(M^c|E,J)P(A|Z)P(Some\,Acute\,Event|M^c)P(Acute\,Event_i),$$

where $P(Z)$ is the probability of choosing agent $j$ from postal code $Z$, $P(E,J|Z)$ is the conditional probability that the agent is of ethnicity $E$ and job class $J$ given postal code $Z$, $P(M|E,J)$ is the conditional probability of the agent being a misuser of alcohol given he is of ethnicity $E$ and job class $J$, $P(A|Z)$ is the probability of alcohol availability $A$ in postal code $Z$, $P(Some\,Acute\,Event|M)$ is the probability that the agent will create some acute event given that he is a misuser of alcohol, and, finally, $P(Acute\,Event_i)$ is the unconditional probability that Acute Event $i$ occurs (in a given day). Of course, $M^c$ indicates the agent is not a misuser of alcohol.

The model simulates an average day for each postal code and for each ethnicity and job class. In the general model, the complexity has increased substantially because population classes will be further split by gender and age categories. Also, instead of an "average" day, we estimate weekend and non-weekend days separately. The model was implemented as a web application. This simulation is available at http://alcoholecology.com. This website presents a JAVA-based implementation and allows the user to explore adjustments in alcohol availability and ethnic balance for the whole county and within individual postal codes.

| Outcome Type | Actual/Year | Simulated/Year | Mean | MSE |
|---|---|---|---|---|
| DWI | 722 | 658 | 708 | 19.6 |
| Assault | 133 | 107 | 132 | 2.0 |
| Murder | 6 | 4 | 7 | 1.0 |
| Sexual Assault | 32 | 38 | 34 | 4.1 |
| Domestic Violence | 161 | 168 | 168 | 16.1 |
| Child Abuse | 221 | 213 | 216 | 17.4 |
| Suicide | 97 | 84 | 100 | 2.6 |
| Benign | 998993 | 998728 | 998635 | 888.4 |

Table 1: Actual and Simulated Results for the Model

As can be seen from Table 1, even with fairly simple assumptions, the directed graph model reasonably approximates the actual data. The column labeled "Mean" is calculated by running the simulation 100 times and transforming this estimate to a yearly average value. The "MSE" column represents the error associated with the simulation results when compared with the actual values using the probabilities for the year of the actual values.

# 4 Conclusions

Alcohol studies have traditionally focused on specific acute outcomes and intervention strategies for mitigating those particular acute effects. Such approaches often have unintended consequences so that intervening to reduce one acute effect may increase other undesirable outcomes. Geospatial and temporal effects have been considered separately, but are not traditionally integrated with intervention models. Statistical analysis in alcohol studies is done often at a relatively elementary level applying relatively elementary hypothesis testing procedures. Indeed, it is difficult to find outlets for modeling innovations within the traditional alcohol literature. In the present paper, we outline a comprehensive modeling framework based on agent-based simulation, social networks, and directed graphs that captures some of the complexity of the alcohol ecological system. We aggregate agents into relatively homogeneous clusters, but provide diversity by stochastic modeling of outcomes. We provide a framework for temporal and geospatial effects and discuss proxies for approximating these effects. We calibrate the model, again using approximations based on actual data.

# Acknowledgements

The work of Dr. Wegman is supported in part by the U.S. Army Research Office under contract W911NF-04-1-0447. The work of Dr. Said is supported in part by Grant Number F32AA015876 from the National Institute on Alcohol Abuse and Alcoholism. The content is solely the responsibility of the authors and does not necessarily represent the official views of the National Institute on Alcohol Abuse and Alcoholism or the National Institutes of Health.

# References

1. Collins JJ, Schlenger WE (1988) Acute and chronic effects of alcohol use on violence. Journal of Studies on Alcohol 49(6):516-521
2. Ezatti M, Lopez A, Rodgers A, Vander Hoorn S, Murray C (2002) Comparative risk assessment collaborating group. Selected major risk factors and global and regional burden of disease. Lancet 360:1347-1360
3. Gorman DM, Speer PW, Gruenewald PG, Labouvie EW (2001) Spatial dynamics of alcohol availability, neighborhood structure and violent crime. Journal of Studies on Alcohol 62:628-636
4. Gruenewald PJ, Remer L, Lipton R (2002) Evaluating the alcohol environment: Community geography and alcohol problems. Alcohol Research and Health 26
5. Leonard KE, Collins RL, Quigley BM (2003) Alcohol consumption and the occurrence and severity of aggression: An event-based analysis of male-to-male barroom violence. Aggressive Behavior 29:346-365

6. National Institute on Alcohol Abuse and Alcoholism (1998) Drinking in the United States: Main findings from the 1992 national longitudinal alcohol epidemiologic survey (NLAES). NIH Publication No. 99-3519. U.S. Alcohol Epidemiologic Data Reference Manual. 6.
7. Said YH, Wegman EJ (2007) Quantitative assessments of alcohol-related outcomes. Chance 20(3):17-25

# Using Participatory Learning to Model Human Behavior

Ronald R. Yager

yager@panix.com, Machine Intelligence Institute, Iona College, New Rochelle, NY

**Abstract** We discuss the need for addressing the issue of learning in human behavior modeling and social computing. We introduce and describe the participatory learning paradigm. A formal system implementing this type of learning agent is provided. We then extend this system so that it can learn from interval type observations.

## 1 Introduction

Any attempt at human behavioral modeling and social computing must include the activity of human learning. At a micro level building computer-based human like synthetic agents requires providing them with suitable learning mechanisms. At a more global level, many human social interactions are managed by ones understanding of others learning mechanism. A sophisticated understanding of the learning process can play an important role in helping one achieve their goals with other human beings. These goals can be overt and benevolent ones such as teaching and explaining. They can covert and less benevolent such as trying to convince another person to accept your position or beliefs. The modern media makes an extensive use of an understanding of the mechanism of human learning in order to convey its messages and sell its products. Clearly successful politicians as well as elected government officials relay on their ability to explain to us what is correct as they see it. Many of the courtroom activities engaged in by a lawyer benefit by an understanding and use of learning theory. A lawyer's interaction with a jury on behalf of his client clearly can benefit from an understanding of how people on the jury process information. A judge's success in explaining to a jury a complex legal issue can be enhanced by a deeper understanding of the human learning process.

Providing learning mechanisms for our computational systems is a major interest of modern information based technology. The discipline of computational learning theory deals exclusively with this problem. Neural network technology also has a major interest in learning theory.

The development of computational learning techniques involves a complex interaction between various criteria. On one hand we are attempting to the model human learning mechanisms which have been developed over man's long history and proven

H. Liu et al. (eds.), *Social Computing and Behavioral Modeling*,
DOI: 10.1007/978-1-4419-0056-2_31, © Springer Science + Business Media, LLC 2009

their worth by enduring Darwin's survival of the fittest test. On the other hand are the constraints of implementing these biological mechanisms with digital operations. A further consideration is how much use should be made of the ability of computers to rapidly process large amounts of information, an ability lacking in humans.

In this work we shall describe a paradigm of learning called participatory learning [1-3]. We believe that this paradigm can help support the digital modeling of some interesting aspects of human learning in synthetic agents as well provide real humans with a deeper understanding of the learning process.

Much of the information available to human being is expressed using words. As noted by Zadeh [4, 5] a central feature of this type of information is its lack of precision, it has a granular nature. The discipline of granular computing [6] has been emerging to provide tools to represent this type of information. Here we consider the application of the participatory learning paradigm to some situation in which our observations are granules.

## 2 The Participatory Learning Paradigm

As noted by Quine [7, 8] human leaning is not context free as it is pervasively effected by the belief state of the agent doing the learning. To help include this reality in computational learning schemes we introduced the paradigm of participatory learning [1]. The basic premise of the participatory learning paradigm is that learning takes place in the framework of what is already learned and believed. The implication of this is that every aspect of the learning process is affected and guided by the current belief system. Our choice of name - participatory learning - is meant to highlight the fact that when learning we are in an environment in which our current knowledge participates in the process of learning about itself. The now classic work by Kuhn [9] describes related ideas in the framework of a scientific advancement.

Figure 1 Partial view of participatory learning process

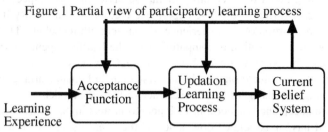

In figure 1 we provide a partial view of a prototypical participatory learning process which highlights the enhanced role played by the current belief system. Central to the participatory learning paradigm is the idea that observations too conflicting with our current beliefs are generally discounted. An experience presented to the system is first sent to the acceptance or censor component. This component, which is under the control of the current belief state, decides whether the experience is compatible with the current state of belief, if it is deemed as being compatible the experience is passed along to the learning component which uses this experience to update the current belief. If the experience is deemed as being too incompatible it is rejected and not used for learning. Thus we see that the acceptance component acts as a kind of filter with respect to deciding which experiences are to be used for learning. We emphasize here that the state of the current beliefs participates in this filtering operation. We note that many learning paradigms do not include a filtering mechanism of the type provided by the acceptance function and let all data pass through to modify the current belief state. Often in these systems stability is obtained by using slow learning rates. Participatory learning has the characteristic of protecting our belief structures from wide swings due to erroneous and anomalous observations while still allowing the learning of new knowledge.

Because of the above structure a central characteristic of the PLP (**Participatory Learning Paradigm**) is that an experience has the greatest impact in causing learning or belief revision when it is compatible with our current belief system. In particular, observations too conflicting with our current beliefs are discounted. As shown in [1] the rate of learning using the PLP is optimized for situations in which we are just trying to change a small part of our current belief system. The structure of the participatory learning system (PLS) is such that it is most receptive to learning when confronted with experiences that convey the message "what you know is correct except for this little part." On the other hand a PLS when confronted with an experience that says "you are all wrong, this is the truth" responds by discounting what is being told to it. In its nature, it is a conservative learning system and hence very stable. We can see that the participatory learning environment uses sympathetic experiences to modify itself. Unsympathetic observations are discounted as being erroneous. Generally a system based on the PLP uses the whole context of an observation (experience) to judge something about the credibility of the observation with respect to the learning agent's beliefs, if it finds the whole experience credible it can modify its belief to accommodate any portion of the experience in conflict with its belief. That is if most of an experience or observation is compatible with the learning agent's current belief the agent can use the portion of the observation that deviates from its current belief to learn. It should be pointed out that in this system as in human learning and most other formal learning models the order of experiences effects the model.

While the acceptance function in PLP acts to protect an agent from responding too "bad" data it has an associated downside. If the agent using a PLP has an incorrect belief system about the world it constrains this agent to remain in this state of blissful ignorance by blocking out correct observations that may conflict with this erroneous belief model. In figure 2 we provide a more fully developed version of the participatory learning paradigm which addresses this issue by introducing an arousal mechanism in its guise of a critic.

Figure 2.Fully developed participatory learning process

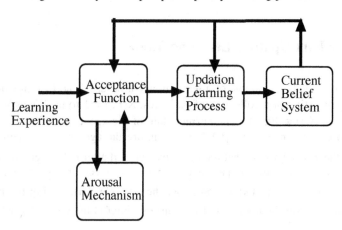

The role of the arousal mechanism, which is an autonomous component not under the control of the current belief state, is to observe the performance of the acceptance function. In particular, if too many observations are rejected as being incompatible with the agent's belief model, the learning agent becomes aroused that something may be wrong with its current state of belief, and a loss of confidence is incurred. The effect of this loss of confidence is to weaken the filtering aspect of the acceptance component and let incoming experiences that are not necessarily compatible with the current state of belief to be used to help update the current belief. This situation can result in a rapid learning in the case of a changing environment once the agent has been aroused. Essentially the role of the arousal mechanism is to help the agent to get out of a state of belief that is deemed as false.

Fundamentally we see two collaborating mechanisms at play in this participatory learning paradigm. The primary mechanism manifested by the acceptance function and controlled by the current state of belief is a conservative one, it assumes the current state of belief is substantially correct and only requires slight tuning. It rejects strongly incompatible experiences and doesn't allow them to modify its current belief. This mechanism manifests its effect on each individual learning experience. The secondary mechanism, controlled by the arousal mechanism, being less conservative allows for the possibility that the agent's current state of belief may be wrong. This secondary mechanism is generally kept dormant unless activated by being aroused by an accumulation of input observations in conflict with the current belief. What must be emphasized is that the arousal mechanism contains no knowledge of the current beliefs, all knowledge resides in the belief system, it is basically a scoring system calculating how the current belief system is performing. It essentially does this by noting how often the system has encountered incompatible observations. Its effect is not manifested by an individual incompatible experience but by an accumulation of these.

# 3 A Basic Participatory Learning Model

In [1] we provided an example of a learning agent based of the PLP in which we have a context consisting of a collection of variables, $X(i)$, $i = 1$ to n. We assumed that the values of the $X(i) \in [0, 1]$. The current state of the agents belief consist of a vector $\mathbf{V_{k-1}}$ whose components, $V_{k-1}(i)$, $i = 1$ to n, are the agents current belief about the values of the $X(i)$. This is what the agent has learned after k-1 observations. The current observation (learning experience) consists of a vector $\mathbf{D_k}$, whose component $d_k(i)$, $i = 1$ to n, is an observation about the variable $X(i)$. Using the participatory learning mechanism the updation of our current belief is the vector $\mathbf{V_k}$ whose components are

$$V_k(i) = V_{k-1}(i) + \alpha \rho_k^{(1 - a_k)}(d_k(i) - V_{k-1}(i)) \qquad \textbf{(I)}$$

Using vector notation we can express this as

$$\mathbf{V_k} = \mathbf{V_{k-1}} + \alpha \rho_k^{(1 - a_k)}(\mathbf{D_k} - \mathbf{V_{k-1}}) = \alpha \rho_k^{(1 - a_k)}\mathbf{D_k} + (1 - \alpha \rho_k^{(1 - a_k)})\mathbf{V_{k-1}}$$

In the above $\alpha \in [0, 1]$ is the basic learning rate. We see $\alpha$ functions like the learning rate in most updation algorithms, the larger $\alpha$ the more the update value is effected by the current observation. The term $\rho_k$ is the compatibility of the observation $\mathbf{D_k}$ with the current belief $\mathbf{V_{k-1}}$. In [1] it was suggested that we calculate

$$\rho_k = 1 - \frac{1}{n}\sum_{i=1}^{n} |d_k(i) - V_{k-1}(i)| \qquad \textbf{(II)}$$

It is noted that $\rho_k \in [0, 1]$. The larger $\rho_k$ the more compatible the observation is with the current belief. The agent looks at the individual compatibilities as expressed by $|d_k(i) - V_{k-1}(i)|$ to determine the overall compatibility $\rho_k$. Here we see that if $\rho_k$ is small, the observation is not compatible, then we tend to filter out the current observation. On the other hand if $\rho_k \rightarrow 1$, then the current observation is in agreement with the belief and it is allowed to affect the belief. We refer to observation with large $\rho_k$ as "kindred" observations.

The term $a_k$, also lying in the unit interval, is called the arousal rate. It is inversely related to the confidence in the model based upon its past performance. The smaller

$a_k$ the more confident. The larger the arousal rate the more suspect we are about the credibility of our current beliefs. The calculation for $a_k$ suggested in [1] is

$$a_k = (1 - \beta) a_{k-1} + \beta (1 - \rho_k) \quad \textbf{(III)}$$

Here $\beta \in [0, 1]$ is a learning rate. $\beta$ is generally less than $\alpha$.

While here we shall restrict the values of the $X(i)$ to be the unit interval this is not necessary. We can draw the $X(i)$ from some subset $S_i$ of the real line . However, in this case we must use a proximity relationship [10], $\text{Prox}_i: S_i \times S_i \rightarrow [0, 1]$ such that $\text{Prox}(x, x) = 1$ and $\text{Prox}(x, y) = \text{Prox}(y, x)$. More generally we need some methodology for taking an observation $\mathbf{D_k}$ and the current belief $\mathbf{V_{k-1}}$ and determining $\text{Comp}(\mathbf{D_k}, \mathbf{V_{k-1}})$ as a value in the unit interval.

## 4 Learning From Interval Observations

We shall now consider the issue of learning when our observations can be granular objects [6, 11]. Here then we shall assume a collection of n variables $X(i)$, $i = 1$ to $n$. Again for simplicity we shall assume that each $X(i)$ takes its value in the unit interval. We note as pointed out above that more generally $X(i)$ can be drawn from space $S_i$ on which there exists some proximity relationship, $\text{Prox}_i(x, y) \in [0, 1]$ for x and y $\in S_i$. Here an observation $\mathbf{D_k}$ has components $d_k(i)$ which are observations about the variable $X(i)$. Here, however we shall allow $d_k(i)$ to be granular objects.

We shall use the basic participatory learning algorithm

$$V_k(i) = V_{k-1}(i) + \alpha \rho_k^{(1 - a_k)}(d_k(i) - V_{k-1}(i))$$

where $\rho_k = 1 - \dfrac{1}{n} \sum_{i=1}^{n} | d_k(i) - V_{k-1}(i) |$ and $a_k = \bar{\beta} a_{k-1} + \beta \bar{\rho}_k$ .

Initially we shall consider the case where while the observations are granular our desire is to learn precise values for the variables $X(i)$. We shall begin our investigation by further assuming that the granular observations are intervals. Thus here our observations are $d_k(i) = [L_k(i), U_k(i)]$ where $L_k(i)$ and $U_k(i)$ are the lower and upper ends of the interval. It is clear that if $L_k(i) = U_k(i)$ then we have a point observation.

Let us first consider the calculation of $V_k(i)$ in this case. Here we shall assume that we have already calculated $\rho_k$ and $a_k$. Using these values we calculate $\alpha \rho_k^{(1 - a_k)} = \delta_k$. Now we must calculate $V_k(i) = V_{k-1}(i) + \delta_k (d_k(i) - V_{k-1}(i))$ where $d_k(i) = [L_k(i), U_k(i)]$. Using interval arithmetic we get that

$$d_k(i) - V_{k-1}(i) = [L_k(i) - V_{k-1}(i), U_k(i) - V_{k-1}(i)]$$

Since $\delta_k \geq 0$ then

$$\delta_k[L_k(i) - V_{k-1}(i), U_k(i) - V_{k-1}(i)] = [\delta_k(L_k(i) - V_{k-1}(i)), \delta_k(U_k(i) - V_{k-1}(i))]$$

From this we get

$$V_k(i) = [V_{k-1}(i) + \delta_k \ (L_k(i) - V_{k-1}(i)), V_{k-1}(i) + \delta_k \ (U_k(i) - V_{k-1}(i))]$$

Since we desire a precise value for $V_k(i)$ we must combine this interval to a single value. The most natural method is to take the midpoint of this

$$V_k(i) = \frac{2V_{k-1}(i) + \delta_k(L_k(i) + U_k(i) - 2V_{k-1}(i))}{2} = V_{k-1}(i) + \delta_k(M_k(i) - V_{k-1}(i))$$

Here $M_k(i) = \dfrac{L_k(i) + U_k(i)}{2}$, is the midpoint of the observed interval

We now turn to the calculation of $\rho_k$ which we previously assumed was available.

Since $\rho_k = \dfrac{1}{n} \sum_{i=1}^{n} \text{Comp}_i(d_k(i), V_{k-1}(i))$ the value of $\rho_k$ depends on the calculation of the individual $\text{Comp}_i(d_k(i), V_{k-1}(i))$. Here since we have assumed the variable X(i) lies in the unit interval $\text{Comp}_i(d_k(i), V_{k-1}(i)) = 1 - \text{Dist}(d_k(i), V_{k-1}(i))$.

Since $d_k(i) = [L_k(i), U_k(i)]$ is an interval we are faced with the problem of calculating the distance of a point $V_{k-1}(i)$ from an interval.

In the following discussion for the sake of clarity we shall temporarily suppress unnecessary indices and just consider the calculation of the distance between a generic interval [L, U] and a point V all lying in the unit interval. We can associate with the interval [L, U] and V two points, the points in the interval closest and farthest from V. We denote these C and F.

We see that these points are as follows

    1. If V ≤ L then C = L and F = U

    2 If V ≥ U then C = U and F = L

    3 If V ∈ [L, U] and U - V ≥ V - L then C = V and F = U

       If V ∈ [L, U] and (U - V) < V - L then C = V and F = L

Using the values C and F we can calculate D.C = |V - C| and D.F = |V - F| as the closest and furthest distance of V from the observation. We note that if we are using a proximity relationship then we calculate Prox(V, F) and Prox(V, C). Since in unit interval case we can calculate Prox(x, y) = 1 - Dist(x, y) shall use the more general concept of proximity. At this point the learning agent has some degree of freedom in calculating the compatibility, Comp(D, V). In order to model this freedom we introduce a parameter $\lambda \in [0, 1]$ and express $\text{Comp}(D, V) = \lambda \text{Prox}(V, C) + \bar{\lambda} \ \text{Prox}(V, F)$.

If $\lambda = 1$ we use the closest value in D to calculate the compatibility and if $\lambda = 0$ we use the furthest.

Here $\lambda$ provides a parameter by which we can characterize our learning agent. We note that if $\lambda$ is large, closer to one, the agent is more willing to accept new observation and hence more open to learning. On the other hand the smaller $\lambda$, closer to zero, the less open the agent is to using observation to learn. It is more conservative.

We should note that in building agents that learn we have the additional freedom of specifying $\lambda$ globally for all variables or providing a different value $\lambda_i$ for each variable. An additional degree of sophistication can be had by allowing $\lambda$ to change rather than being fixed. In particular $\lambda$ can be a function of the arousal rate.

We note once we have determined $\rho_k$ the calculation of $a_k$ poses no new challenges. However there is an interesting observation we can make. We note that selecting $\lambda$ to be big will result in bigger compatibilities and hence a bigger value for $\rho_k$. This will of course, as we already noted, make the learning of $V_k$ more open to accept the observation and hence make for faster learning. However there is an interesting counter balancing effect. Making $\rho_k$ large has the effect of making the arousal factor $a_k$ smaller. Since the full term used in equation I is $\rho_k^{(1-a_k)}$ a smaller value for $a_k$ tends to slow down the learning because we are assuming the model is good. Selecting $\lambda$ to be small will of course have the opposite, it will diminish $\rho_k$, which tends to reduce the effect of the current observation but on the other hand it will tend to increase $a_k$, the arousal rate, which has the effect of making the system more willing to receive new observations. Thus the effect of $\lambda$ has very interesting dynamics which appears to be a kind of delayed feedback. This effect appears to balancing and doesn't allow the learning to get to extreme.

An interesting possible type of learning agent is one that has two values for the parameter $\lambda$ used to determine the $\rho_k$. Let us denote one as $\lambda_I$ and the other $\lambda_{III}$. Here $\lambda_I$ is used to determine the compatibility and eventually the value of $\rho_k$ used in the formula I. While $\lambda_{III}$ is determined to determine the compatibility value $\rho_k$ used in calculating $a_k$. We see that by making $\lambda_I$ large and $\lambda_{III}$ small we are tending to be extremely open to learning. In this case we tend to believe the current observation is compatible with our beliefs for learning $V_k$. On the other hand we tend to assume that the observation was not compatible for determining our rate of arousal. This tends to make $a_k$ large. In combination these together lead to larger values for $\rho_k^{(1-a_k)}$ making the learning faster.

An opposite effect can be observed if $\lambda_I$ is made small and $\lambda_{III}$ is made large. Here the agent is tending to believe the observation is incompatible with his current beliefs for learning $V_k$. On the other hand he is assuming that the observation is compatible for determining the arousal rate. This tends to make $a_k$ smaller. In combination these together lead to smaller values for $\rho_k^{(1-a_k)}$ tending to block any learning. Here the agent is being very conservative.

Values of $\lambda_I$ and $\lambda_{III}$ very different may be useful to model kinds of neurotic behavior.

## 5 Conclusion

We described the participatory learning paradigm. A notable feature associated with this paradigm is the use of the compatibility between observations and the current belief to modulate the learning. Another feature of this paradigm is the use of two levels of learning, the one, constraint by the current beliefs, dealing with the immediate task of learning from the current observation and the second, independent of the current belief system, observing the performance of the current belief systems. A formal system implementing this type of learning agent was described. This system involved equations for updation, determining the compatibility of the observation with the current belief and calculating the arousal level. We extended this system so that it can learn from interval type observations. We note the participatory learning paradigm can be used in at least two modes. One mode is as framework for providing a learning mechanism when building intelligent agents. A second is a tool to help us understand the role that learning plays in social interactions.

## References

1. Yager, R. R., "A model of participatory learning," IEEE Transactions on Systems, Man and Cybernetics 20, 1229-1234, 1990.
2. Yager, R. R., "Participatory learning: A paradigm for building better digital and human agents," Law, Probability and Risk 3, 133-145, 2004.
3. Yager, R. R., "Extending the participatory learning paradigm to include source credibility," Fuzzy Optimization and Decision Making, (To Appear).

4. Zadeh, L. A., "Toward a perception-based theory of probabilistic reasoning with imprecise probabilities," Journal of Statistical Planning and Inference 105, 233-264, 2002.
5. Zadeh, L. A., "From imprecise to granular probabilities," Fuzzy Sets and Systems, 370-374, 2005.
6. Bargiela, A. and Pedrycz, W., Granular Computing: An Introduction, Kluwer Academic Publishers: Amsterdam, 2003.
7. Quine, M. V. O., "Two dogmas of empiricism," Philosophical Review 60, 20-43, 1951.
8. Quine, M. V. O., From a Logical Point of View, Harvard Press: Cambridge, 1953.
9. Kuhn, T. S., The Structure of Scientific Revolutions, University of Chicago Press, 1962.
10. Kaufmann, A., Introduction to the Theory of Fuzzy Subsets: Volume I, Academic Press: New York, 1975.
11. Lin, T. S., Yao, Y. Y. and Zadeh, L. A., Data Mining, Rough Sets and Granular Computing, Physica-Verlag: Heidelberg, 2002.
12. Janssens, S., De Baets, B. and De Meyer, H., "Bell-type inequalities for quasi-copulas," Fuzzy Sets and Systems 148, 263-278, 2004.
13. Alsina, C., Trillas, E. and Valverde, L., "On some logical connectives for fuzzy set theory," J. Math Anal. & Appl. 93, 15-26, 1983.
14. Klement, E. P., Mesiar, R. and Pap, E., Triangular Norms, Kluwer Academic Publishers: Dordrecht, 2000.
15. Yager, R. R., "Generalized triangular norm and conorm aggregation operators on ordinal spaces," International Journal of General Systems 32, 475-490, 2003.

# Biographies

## Huan Liu, ASU

Huan Liu is on the faculty of Computer Science and Engineering at Arizona State University. He received his Ph.D. and MS. (Computer Science) from University of Southern California, and his B.Eng. (Computer Science and Electrical Engineering) from Shanghai Jiaotong University. His research interests include machine learning, data and web mining, social computing, and artificial intelligence, investigating search and optimization problems that arise in many real-world applications with high-dimensional data such as text categorization, biomarker identification, group profiling, searching for influential bloggers, and information integration. He publishes extensively in his research areas, including books, book chapters, encyclopedia entries, as well as conference and journal papers.

## John J. Salerno, AFRL

John Salerno has been employed at the Air Force Research Laboratory for the past 29 years where he has been involved in many research disciplines. He began his career in the Communications Division where he was involved in and ran a number of research efforts in both voice and data communications networks. In 1989 he moved from the Communications Division into the Intelligence and Exploitation Division. Since then he has been involved in research in distributed heterogeneous data access architectures, web interfaces, and in the further definition and implementation of fusion to various application areas. He has designed, built and fielded a number of operational systems for the Intelligence Community. He holds two US patents and has over 50 publications. John received his PhD from Binghamton University (Computer Science), a MSEE from Syracuse University, his BA (Mathematics and Physics) from SUNY at Potsdam and an AAS (Mathematics) from Mohawk Valley Community College. He is an AFRL Fellow.

**Michael J. Young, AFRL**

Michael J. Young is Chief of the Behavior Modeling Branch of the Air Force Research Laboratory where he directs a variety of research on human factors and socio-cultural modeling. During his twenty-two year career his primary research goal has been improving the representations of human behavior in training systems, command post exercises, and decision aids. His current research explores the incorporation of cultural factors into computational models and the development of mathematical approaches for defining cultural groups and understanding how they evolve over time. He received a PhD from Miami University (Experimental Psychology), and an M.A. (Psychobiology) and A.B. (Psychology) from the University of Michigan.

# Author Index

George Cybenko, 180

**D**

Gary Danielson, 190

**E**

Mark A. Ehlen, 76

**F**

Anthony J. Ford, 85

**G**

Norman Geddes, 155
Charles Gieseler, 42
Vadas Gintautas, 93
Matthew Glickman, 42
Rebecca Goolsby, 2

**H**

Aric Hagberg, 93
Michael Hechter, 102
Tristan Henderson, 16
Shoji Hirano, 219
Shuyuan Mary Ho, 113
Lee Hoffer, 50

**J**

Ruben Juarez, 123

## Z